INVERSE AND ILL-POSED PROBLEMS SERIES

Computer Modelling in Tomography and Ill-Posed Problems

Also available in the Inverse and Ill-Posed Problems Series:

An Introduction to Identification Problems via Functional Analysis
A. Lorenzi
Coefficient Inverse Problems for Parabolic Type Equations and Their Application
P.G. Danilaev
Inverse Problems for Kinetic and other Evolution Equations
Yu.E. Anikonov
Inverse Problems of Wave Processes
A.S. Blagoveshchenskii
Uniqueness Problems for Degenerating Equations and Nonclassical Problems
S.P. Shishatskii, A. Asanov and E.R. Atamanov
Uniqueness Questions in Reconstruction of Multidimensional Tomography-Type Projection Data
V.P. Golubyatnikov
Monte Carlo Method for Solving Inverse Problems of Radiation Transfer
V.S. Antyufeev
Introduction to the Theory of Inverse Problems
A.L. Bukhgeim
Identification Problems of Wave Phenomena - Theory and Numerics
S.I. Kabanikhin and A. Lorenzi
Inverse Problems of Electromagnetic Geophysical Fields
P.S. Martyshko
Composite Type Equations and Inverse Problems
A.I. Kozhanov
Inverse Problems of Vibrational Spectroscopy
A.G. Yagola, I.V. Kochikov, G.M. Kuramshina and Yu.A. Pentin
Elements of the Theory of Inverse Problems
A.M. Denisov
Volterra Equations and Inverse Problems
A.L. Bughgeim
Small Parameter Method in Multidimensional Inverse Problems
A.S. Barashkov
Regularization, Uniqueness and Existence of Volterra Equations of the First Kind
A. Asanov
Methods for Solution of Nonlinear Operator Equations
V.P. Tanana
Inverse and Ill-Posed Sources Problems
Yu.E. Anikonov, B.A. Bubnov and G.N. Erokhin
Methods for Solving Operator Equations
V.P. Tanana
Nonclassical and Inverse Problems for Pseudoparabolic Equations
A. Asanov and E.R. Atamanov
Formulas in Inverse and Ill-Posed Problems
Yu.E. Anikonov
Inverse Logarithmic Potential Problem
V.G. Cherednichenko
Multidimensional Inverse and Ill-Posed Problems for Differential Equations
Yu.E. Anikonov
Ill-Posed Problems with A Priori Information
V.V. Vasin and A.L. Ageev
Integral Geometry of Tensor Fields
V.A. Sharafutdinov
Inverse Problems for Maxwell's Equations
V.G. Romanov and S.I. Kabanikhin

INVERSE AND ILL-POSED PROBLEMS SERIES

Computer Modelling in Tomography and Ill-Posed Problems

M.M. Lavrent'ev, S.M. Zerkal and O.E. Trofimov

///VSP///
UTRECHT • BOSTON • KÖLN • TOKYO
2001

VSP
P.O. Box 346
3700 AH Zeist
The Netherlands

Tel: +31 30 692 5790
Fax: +31 30 693 2081
vsppub@compuserve.com
www.vsppub.com

First published in 2001

ISBN 90-6764-350-5

Printed in The Netherlands by Ridderprint bv, Ridderkerk.

Preface

The theory of ill-posed problems is the basis of mathematical apparatus of tomography formed to this moment as important intensively developed direction in computer technologies. Classical tomography using the inversion of integral transformations and required complete projection data is researched sufficiently well and is widely applied in various domains of human activity. The comparatively weakly researched untraditional tomography problems are solving thank to the new achievements in calculation mathematics and the theory of ill-posed problems, the regularization process of solving ill-posed problem, and the increase of stability. Computerized tomography is the applied direction in science, and its development denotes mathematical modelling and computer experiments. These experiments show us the possibilities and applicability of algorithms of processing tomography data. This monograph is devoted to considering these problems in connection with series of ill-posed problems in tomography setting arising in practice.

The theory and experimental material on this subject (both physical and calculating) is sufficiently various and wide. This is explained by applying the methods of computerized tomography into different domains of science and techniques. As the problems of computerized tomography are so wide the authors restricted theirselves by account of the material relating to the domains of their own interests. There are no sections "Conclusion" and "Epilogue" in this book; the corresponding material is included by the authors to "Introduction". Besides, the list of literature is composed so that not all present sources have references in the text, the preference was given to earlier works.

We thank the postgraduate student O. N. Belousova and the scientific workers of the Institute of Automatics and Electrometry of Siberian Branch of Russian Academy of Sciences D. V. Badazhkov and C. N. Kasjanova who had fulfilled the large redactor and designing work in the creation of the manuscript.

The authors participation in the work was as follows: Chapter 1 was written by M. M. Lavrent'ev and O. E. Trofimov jointly, Chapter 2 — by O. E. Trofimov, and Chapter 3 — by S. M. Zerkal.

M. M. Lavrent'ev

S. M. Zerkal

O. E. Trofimov

Contents

Introduction

Last decades of our century were marked by the appearance of a new field of mathematics — computational tomography. This theory is the basis for solution of many applied problems. The methods of computational (computerized) tomography make it possible to find out the interior structure of a body by examining the characteristics of radiation passing through the object under study (transmission tomography).

Depending on the type of radiation used, we distinguish X-ray, optical, seismic, and some other kinds of tomography. We also distinguish medical and industrial tomography, geotomography, and so on, depending on the field of application. Tomography problems feature inverse causal connection, i.e., we reconstruct the characteristics of the object in question from the effect, namely, from the measured sound signal. As a cause, these characteristics affect the signal parameters. For this reason, such problems are called inverse problems.

Intensive development of computerized tomography is facilitated also by the development of computing technologies, computational mathematics, the theory of inverse problems of mathematical physics, and the theory of ill-posed problems. The main problems of the theory of ill-posed problems that are relevant to this monograph are stated in Chapter 1 (for details, see Lavrent'ev, Romanov, and Shishatskii, 1986).

One of the most important applications of tomography is flaw detection in industry. In the USSR, this kind of application was well-developed in missile industry. It was one of the few types of industry in which USSR was among the leading countries. For this reason, new technologies have been developed including those of detecting small crack-type defects (see Zerkal, 1990). One of such problems is described in Appendix. A variety of physical conditions in practical problems that are solved using computerized tomography has generated a variety of mathematical settings. In its turn, this has brought about the development of the corresponding numerical methods and algorithms, which would have been impossible without the development of mathematical fundamentals of computerized tomography (see Bronnikov and Voskoboinikov, 1989, 1990; Natterer, 1986; Pickalov, 1985; Pickalov and Preobrazhensky, 1987; Tikhonov, Arsenin, and Timonov, 1987). The classical integral transformation of I. Radon is the basis of tomography. Though Radon's classical work was published in 1917, even nowadays this transformation is a subject of both theoretical and practical research for mathematicians. The above-mentioned Radon's work can be found, for example, in the appendix of the monograph of Helgason (1980).

In the mathematical literature devoted to the methods of computerized tomography, the inversion formulas of the Radon transform are usually given in the original form. However, these formulas assume the most complete and general form when employing the mathematical apparatus of generalized functions. It should be noted that Radon's research had been carried out before the theory of generalized functions took final shape.

The apparatus of generalized functions is very useful also for construction of numerical algorithms. In Section 1.4 it is shown that the well-known and widely used algorithm of convolution and inverse projection is constructed on the basis of a sequence of regular functions converging to the generalized function $1/z^2$. In this section we also describe a modification of the algorithm of convolution and inverse projection which involves the canonical regularization of the generalized function $1/z^2$.

Generalized functions are even of greater importance in 3D cone-beam tomography reconstruction.

In the classical computerized X-ray tomography, the process of reconstruction of the density of a three-dimensional object consists of a sequence of reconstructions of a function of two variables which characterizes the object density in the corresponding sequence of two-dimensional sections of the object. At each step we solve the problem of inverting the two-dimensional Radon transform.

The scheme of three-dimensional cone-beam tomography reconstruction is a generalization of the classical scheme. In the three-dimensional scheme, the density is reconstructed at every point from the values of integrals along a certain set of straight lines in the three-dimensional space. In the case of the classical computerized tomography, the corresponding set consists of all the lines lying in the sections orthogonal to one of the coordinate axes. The following scheme is often used for cone-beam tomography. The source of radiation is moving along a certain spatial curve. In this case, the mathematical problem is to reconstruct a function of three variables from the integrals along the straight lines passing through the given curve. This problem was studied for various classes of functions and curves (see Anikonov, 1978; Gel'fand and Goncharov, 1987; Denisyuk, 1991; Kirillov, 1961; Orlov, 1975; Tuy, 1983). In Tuy (1983) the formula of inversion for functions with finite support and for the curves satisfying certain conditions was obtained. The main condition was that each plane intersecting the object intersects (transversely) the trajectory of the source. An example of such a curve which satisfies the condition of Tuy (1983) is a spiral. Another example is the complex of two unit circles lying in the intersecting planes. The inversion formula in Tuy (1983) involves generalized functions. This makes it difficult to apply this formula for numerical algorithms (see Natterer, 1986).

In Section 2.1, the inversion formulas for the ray transformation are reduced to the form which involves regular functions and operations on them. Such form of the formulas allows us to construct numerical algorithms for three-dimensional cone-beam tomography reconstruction.

In the X-ray computerized tomography, it is often required to know the density only in a certain subdomain. A typical example is the quality control for the longitudinal welded seams in pipes. We are interested in the reconstruction of the density only in the seam area.

It is well known that the problem of inverting the two-dimensional Radon transform is not a local problem. To calculate the density at a certain point, we need to know the information on all the rays intersecting the object, not only the rays which intersect a neighborhood of the specified point. In contrast to the two-dimensional space, the problem of inverting the Radon transform in the three-dimensional space is local. However, being represented by the integrals over the planes in the three-dimensional space, the initial data of X-ray tomography corresponds to the ray transformation rather than the Radon transform.

As was noted above, the essential conditions are the Kirillov – Tuy conditions (the K.-T. conditions) in the problems of inverting the ray transformation. These conditions are local in the sense that at the points where the K.-T. conditions hold the density can be reconstructed, and at some points the K.-T. conditions may fail.

In particular, when considering welded seams in pipes, we may confine ourselves only to the parts of trajectories that are necessary for the region under consideration.

Note that the K.-T. conditions are not necessary for inverting the ray transformation (see Anikonov, 1973; Blagoveshchenskii, 1986; Finch 1985).

In particular, to reconstruct the density of the object in the three-dimensional space it is sufficient for the object trajectory to consist of only one circle. In this case, the problem will be unstable in the whole (see Finch, 1985); however, we may select a stable part in it. (see Blagoveshchenskii, 1986; Trofimov, 1992). This part consists in calculating the integrals along the straight lines intersecting the disk from the integrals along the straight lines intersecting the circle.

The corresponding algorithms may be used in flaw detection with restrictions imposed on possible directions of rays.

In Section 2.1 the results of computer simulation of the situations mentioned above are represented.

The algorithms of inverting the ray transformation may be normally used in the cases where objects in question, or their defects, have complicated spatial form (spirals, cracks, and so on).

Sections 2.2 and 2.3 are supplementary to the main contents of Chapter 2. In Section 2.2, elements of the theory of generalized functions are applied to the problems of inversion of the Radon transform and the ray transformation.

Section 2.3, some relations between the Radon transform, the Fourier transform, and the ray transformation are presented.

The method of computerized tomography was discussed at regular symposiums in the USSR. The section on geotomography was organized at the III Russian symposium on computerized tomography. This marked the beginning of the development of geotomography in Russia (see Problems of Geotomography, 1997).

There are many publications on this subject. In particular, we should mention the works of Dziewonski *et al.*, which contain the description of interesting global Earth models constructed on the basis of kinematic data. However, the degree of reliability of these results is not high. (see Seismic Tomography, 1987, and the bibliography therein).

Nowadays, the term "seismic geotomography" is generally accepted in geophysical terminology. This is caused by interesting and promising results in solving the inverse kinematic problem, which is one of the geophysical problems being under most intensive study. It is also applied in optics and acoustics. In mathematical setting, it is reduced to the problem of integral geometry and is connected with variational calculus. The inverse kinematic problem is a typical example of a geophysical problem whose different formulations are classical for computerized tomography. In this problem it is required to determine the velocity characteristics of a medium from the time of the waves passing along the geodesic curves (the rays) from the sources to the receivers. This problem is a practical example of the problem of integral geometry, whose general statement is presented in Section 1.2. Herglotz and Weichert (1907) were the first to obtain the results of solving the inverse kinematic problem of seismology for spherically symmetric Earth. Further investigations were focused on solving the one-dimensional problem. In particular, the ambiguity of the solution of this problem was studied by Gerver and Markushevich. It was only in sixties, beginning with the works of Lavrent'ev and Romanov, when systematic study of the multidimensional inverse kinematic problem began. The first result (Lavrent'ev and Romanov) for the inhomogeneous inverse kinematic problem was obtained in the case of two-dimensional space x_1, x_2 for the domain $x_2 \geq 0$ in the linearized formulation. The speed of signal in the medium, $V(x)$, was represented as $V(x) = V_0(x) + V_1(x)$, where $V_0(x)$ is known and $V_1(x)$ is unknown and small in comparison with $V_0(x)$. Then, to within the accuracy of $O(V_1^2(x))$, we have

$$\tau(x^0, x) - \tau_0(x^0, x) \approx \int\limits_{\Gamma_0(x^0,x)} \frac{V_1(x)}{V_0^2(x)} |\mathrm{d}x|. \tag{1}$$

Here $\tau(x^0, x)$ is the time of the signal travel from the point x^0 to the point x in the medium under study; $\tau_0(x^0, x)$ is the signal travel time for the medium characterized by $V(x) = V_0(x)$; Γ_0 is the ray trajectory for this medium. For the case

$$\begin{aligned} &V_0(x) = Ax_2 + B \\ &A > 0, \quad B > 0, \quad A = \text{const}, \quad B = \text{const} \end{aligned} \tag{2}$$

it was shown that if A and B are known, the continuous function $V_1(x)$ is determined uniquely in the domain $x_2 \geq 0$ if we know $\tau(x^0, x)$ for each pair of points x^0, x of the straight line $x_2 = 0$. In this case, the problem of determining $V_1(x)$ from the difference $\tau - \tau_0$ from equation (1) has similar character of ill-posedness as the Cauchy problem for the Laplace equation. This means that this problem is strongly ill-posed (see Lavrent'ev, Romanov, and Shishatskii, 1986).

More recently, for a three-dimensional ball and the function $V_0(x)$ of more general form (under certain conditions providing the uniqueness of the solution), V. G. Romanov showed that the solution of equation (1) splits into a set of planes in every section of the large Earth disk (see Romanov, 1987).

A great number of works are devoted to various aspects of the inverse kinematic problem, which is demonstrative of substantial interest in this problem (see Alekseev *et al.*, 1969; Anikonov, 1973; Belolipetskii and Golov, 1989; Belonosova and Alekseev, 1967; Beil'kin, 1978; Gol'din, 1997; Lagunova and Omel'chenko, 1981; Mishen'kina, Shelud'ko, and Krylov, 1983a; Mukhometov, 1975; Seismic Tomography, 1987). Linearized formulations have been developed most intensively. They make it possible to avoid (in part) the difficulty of solving the problem in the general statements. By now there is no constructive solution of the three-dimensional inverse kinematic problem in the general formulation (see Alekseev and Tsibul'chik, 1985). For the most part, numerical algorithms are intended for solution of one-dimensional or strongly ill-posed two-dimensional problems. Determination of the velocity characteristics of the medium is possible not only on the basis of kinematic data; however, considerable progress in this respect is achievable today due to the kinematic approach (see Alekseev and Tsibul'chik, 1985; Puzyrev, 1992).

Numerical simulation experiments for solving the inverse kinematic problem are impossible without simulation of solutuion of two-point direct problems. Furthermore, the study of the direct problem provides better understanding of the inverse problem. The direct kinematic problem in a formulation convenient for the inverse problem in question is solved in Section 3.1.

The direct kinematic problem is considered as the problem of variational calculus. For a medium with linear dependence of the signal propagation speed on the depth, explicit formulas are deduced for the ray and the time of the signal travel along the ray depending on the coordinates of the source and the receiver.

The algorithm of solving an initial boundary value problem is presented. The problem consists in constructing the ray and calculating the time of the signal travel along this ray. It is supposed that the speed distribution for the signal in the media is specified in the analytic form. The algorithm is based on repeated solution of the system of differential equations of the ray by the Runge-Kutta method. The system is extended by adding the equation for the travel time with the initial data that are chosen successively, according to the "adjustment" method. The investigations presented in Sections 3.2–3.3 allow obtaining the new solution of the three-dimensional inverse kinematic problem in the linearized formulation. The observation system characteristic of the method of computerized tomography (circular disposition of the sources and receivers) made it possible to avoid the overdetermination ($\tau(x_0, x)|_{x_3=0}$ is a function dependent on four variables). Also the strong ill-posedness was weakened and the problem became weakly ill-posed (the problem of differentiating a function represented in tabular form).

Using V_0 in the form of a linear function, as was done by Lavrent'ev and Romanov, provides rays Γ_0 in the form of arcs of the circle passing through the points x_0, x, with centre in the line $x_3 = -B/A$. In this case, the rays Γ_0 form the surface of the ball segments based on the circle of the system of observation. Since the projections of the ray onto the plane are segments of straight lines, namely, the chords connecting the points of the circle, the problem of determining the function $n_1(x) = V^{-1}(x) - V_0^{-1}(x)$ on the surface of the ball segment mentioned above is reduced to inversion of the integral Radon transform in the part of the plane $x_3 = 0$ lying within the circle of the observation system (see Bukhgeim, Zerkal, and Pickalov, 1983; Zerkal, 1988). Layer-by-layer reconstruction of the solution of the inverse problem is carried out by means of changing the radius of the circle. As a result, we obtain the family of embedded ball segments filling the three-dimensional domain in question.

In Section 3.3 a numerical algorithm for the inversion formula is presented together with methods of calculating the inverse integral Radon transform (which is the basis of the inversion formula).

The initial data for solution of the inverse problem are calculated using the formulas and algorithms given in Section 3.1. According to the method of computerized tomography, we construct the projection matrix with columns represented by the vectors of the measured data. They are obtained for a fixed observation system which is rotating while the projection matrix is forming around its centre (in the plane $z = 0$) with certain angular step dependent on the chosen number of directions from which the object is illuminated.

In Section 3.4, we discuss the numerical results of the simulation experiments in solution of the inverse kinematic problem for various media which make it possible to analyze the behavior of the solution of the inverse problem depending on the specific features of the initial speed model. Also presented is the numerical study of the stability of the algorithm with respect to random errors of the vector of the initial data for the solution of the inverse problem. The study of the linearization applicability criteria is of independent interest, i.e., determination of the maximum possible difference between the main part of the refraction index and the three-dimensional component. According to the numerical results, the maximum difference may be up to 20% of the main part of the refraction index.

The parameters of the observation system are the number of source-receiver pairs (in the case of the fan-shaped system, it is the number of receivers), which determine the number of elements in one projection, and the number of projections, or raying aspects, of the object under study. These parameters have crucial influence on the quality of the solution of the inverse problem. We have studied how the observation system parameters influence the solution of the inverse problem on the basis of the numerical experiments. We have also constructed the graphs for the mesh norm of the deviation of the desired function from the real one depending on the number of elements in one projection and on the number of projections.

The inverse problems considered in Chapter 3 are ill-posed; therefore, we need the regularization method to solve them. The fundamentals of this method were presented in the works of Tikhonov, Lavrent'ev, and others (see Bakushinsky, 1983; Bakushinsky and Goncharsky, 1989; Lavrent'ev, Romanov, and Shishatskii, 1986; Tikhonov and Arsenin, 1977). Two cases were studied. In the case of weakly ill-posed problem with complete projection data, simple descriptive regularization was applied. With the use of *a priori* information about the smoothness of the solution and the vector of the measured data, this regularization smooths the data and the solution of the inverse problem by means of cubic splines. In the case of incomplete projection data, a more complicated iterative and descriptive regularization procedure was used.

In Section 3.5, the inverse kinematic problem in three-dimensional linearized formulation was considered as a problem of transmission tomography with incomplete data, which is caused by local opacity of the medium. The medium is supposed to have an opaque domain absorbing the probing signal. Tomography problems with incomplete data are studied in a great number of publications. Different formulations of such problems and various solution algorithms have been proposed (see Bronnikov and Voskoboinikov (1989) and the bibliography therein).

The use of the algebraic form of the Radon transform and further application of iterative algorithms (see Bronnikov and Voskoboinikov, 1989, 1990; Lavrent'ev *et al.*, 1995) is one of the ways to solve the problem with incomplete data.

Inhomogeneous discretization of the angles of data scanning is an important specific feature of the problem with incomplete data. In this case, shaded zones appear near the boundary of the convex opaque domain. Information from these zones comes only along certain directions. For iterative algorithms, this implies the appearance of artifacts of "step" type. To suppress the artifacts, we need special methods involving the combination of the median filtration and the averaging by the filter of running mean value. Furthermore, since the problem of artifacts and the difficulty of reconstruction in the shaded zones depend on the discretization and reconstruction domain and the parameters of the observation system, we need a criterion for choosing the parameters of the observation system.

In Section 3.5, a method of effective solution of the inverse kinematic problem for the media with strongly absorbing inclusions is presented.

We give the formulation of the inverse kinematic problem with data that are incomplete as a result of local opacity in the domain under study. We also obtain the main system of linear algebraic equations by reducing the Radon transform to the algebraic form. The regularizing iterative algorithm of successive control of the rows of the system matrix is used in the calculation procedure. The nonlinear combined algorithm presented here makes it possible to overcome the difficulties arising as a result of iterative solution of this problem. In particular, this algorithm provides suppression of intrinsic noise and random errors in the receiver channel. Recommendations for designing the experiment and determining the optimal parameters of the observation system are given. Fulfillment of these recommendations guarantees the exactness of the solution obtained. The numerical simulation experiment for solution of the problem has been implemented.

The inverse kinematic problem with incomplete data often arises in practice. In mathematical modelling in geophysics, considerable attention is focused on the development of the methods of computer simulation of mineral deposits. Computing technologies for petroleum geology is considered to be of prime importance. Most of the efforts made so far were aimed at investigating the problems of the development of static models of oil reservoirs. However, the problems of prospecting and monitoring hydrocarbons that change their parameters during exploitation are of special interest. The study of the methods of solving this problem is developing intensively (see Puzyrev, 1992; Tikhonov, Arsenin, and Timonov, 1987). As a rule, we can only use remote exploration methods to construct a detailed model describing the spatial structure of a deposit. These are seismic methods, such as seismic tomography, which is one of the advanced methods of seismic research based on the method of computerized tomography. It is necessary to develop more advanced theoretical apparatus as the basis for new methods of solving more complicated practical problems (see Puzyrev, 1992). The problem of mathematical modelling arising in this case is the tomography problem with incomplete projection data, which is caused by restrictions imposed on the observation system or opaque inclusions in the medium. In traditional computerized tomography it is necessary to have the complete projection matrix and therefore it is impossible to solve this problem within the framework of this approach. For this reason, we apply the method of iterative regularization and also recurrent algorithms which make effective use of *a priori* information, which is present in excess in geophysical research. This makes it possible to solve the inverse problems of determining speed distributions in the models of geological media. Furthermore, kinematic seismic tomography is a tool that provides insights into the Earth structure and the dynamics of the Earth formation (see Dobretsov, 1997).

Another example, which is of special interest, is a problem being solved with the use of the algorithm that doesn't involve the complete projection matrix. This is the problem of flaw detection involving the diagnosis of crack-type defects that can be traced only in tangent rays, which is considered in Appendix. As a result, a part of information in the projection matrix is missing. Application of the computational methods with the basic idea of the standard model (it is of decisive importance in flaw detection in industry) allows finding the original effective solution of the problem. The algorithm developed in the present work is the so-called local reconstruction. It belongs to an alternative branch of computerized tomography that deals with only a part of the object to be reconstructed. Furthermore, it is possible to apply other methods of the standard model for other geometric conditions that determine the formation of the projection matrix.

Chapter 1.

Mathematical basis of the method of computerized tomography

1.1. BASIC NOTIONS OF THE THEORY OF ILL-POSED PROBLEMS

The theory of ill-posed problems is a field of mathematics which has developed intensively in the last two decades and is connected with most diverse applied problems: interpreting the readings of many physical instruments of geophysical, geological, and astronomical observations, optimization of control, management and planning, synthesis of automatic systems, etc. Development of the theory of ill-posed problems was facilitated by the advent of modern computer technology.

Various areas of the theory of ill-posed problems can be included in traditional areas of mathematics such as function theory, functional analysis, differential equations, and linear algebra.

1. The concept of a well-posed (correct) statement of a problem of mathematical physics was formulated at the beginning of our century by the famous French mathematician Hadamard. Now this concept is represented in textbooks on equations of mathematical physics or partial differential equations.

A problem of mathematical physics or a boundary value problem for a partial differential equation is called well-posed if the following conditions are satisfied: 1) a solution of the problem exists; 2) the solution of the problem is unique; 3) the solution of the problem depends continuously on the data of the problem.

The conditions just formulated need to be refined. Namely, both the solution and the data of the problem are considered as elements of some function space, and the conditions for a problem to be well-posed are formulated as follows:

A. A solution of the problem exists for all data belonging to some closed subspace in a normed linear space of type C^k, L_p, H_p^l, W_p^l and belongs to this space. The subspace is often either the entire space or a part of the space where a finite collection of linear functionals vanishes.

B. The solution of the problem is unique in a certain space.

C. The infinitesimal variations of the solution in the space of solutions correspond to the infinitesimal variations of the data of the problem in the data space.

Hadamard presented an example of an ill-posed problem for a differential equation which, in his opinion, did not correspond to any real physical formulation. This example was the Cauchy problem for the Laplace equation.

1.1. It turned out that the opinion of Hadamard concerning the Cauchy problem for the Laplace equation and a number of other problems of the same type was erroneous. Problems that are ill-posed in the classical sense were encountered long ago in the mathematical description of physical phenomena. A number of classical results in mathematical analysis and differential equations of the last century and the beginning of this one can be attributed to the theory of ill-posed problems. However, the systematic study of questions of the theory of ill-posed problems began comparatively recently.

In discussing ill-posed problems with specialists in natural sciences, mathematicians repeatedly asserted that considerations of ill-posed problems are devoid of meaning, referring here to the opinion and the example of Hadamard. In the thirties and forties, geophysicists discovered a connection between a problem equivalent to the Cauchy problem for the Laplace equation and certain questions of the interpretation of gravitational and magnetic anomalies. Expert mathematicians evaded discussing these problems, and the attempts to carry out calculations without theoretical foundation were ineffectual. The trial-and-error method used by practicing geophysicists also remained without theoretical basis.

In recent years it was established that the set of physical phenomena whose mathematical description is connected with ill-posed problems is not smaller than the set of phenomena associated with problems which are well-posed in the classical sense.

The method of integral equations is one of the main methods of investigating boundary value problems for equations of mathematical physics. Problems that are well-posed in the classical sense are reduced to the Fredholm integral equation of the second kind or to singular integral equations. The latter, after some transformations, are usually also reduced to equations of the second kind. Ill-posed problems are frequently reduced to integral equations of the first kind relatively easily. In particular, this holds for the two ill-posed problems mentioned above. The integral equation to which the Cauchy problem for the Laplace equation and the Cauchy problem for the heat equation with reverse time are reduced has the following forms:

$$\frac{y}{\pi}\int_{-\infty}^{\infty}\frac{\varphi(\xi)}{y^2+(x-\xi)^2}\,\mathrm{d}\xi = f(x), \tag{1.1.1}$$

$$\frac{1}{2\sqrt{\pi t}}\int_{-\infty}^{\infty}\exp\Big[-\frac{(x-\xi)^2}{4t}\Big]\varphi(\xi)\,\mathrm{d}\xi = f(x). \tag{1.1.2}$$

The concept of an operator equation of the first kind is a natural generalization of the concept of an integral equation of the first kind. An operator equation of the first kind is defined as an equation of the form

$$Ax = f, \tag{1.1.3}$$

where x and f are elements of certain spaces X and F; A is a compact (linear) operator acting in X with range F.

It is known that the operator inverse to a completely continuous operator (in an infinite-dimensional space), if it exists, is not continuous; thus, the problem of solving equation (1.1.3) is ill-posed. It is obvious that the problem of solving (1.1.3) may be well-posed if it is considered for some other pair of spaces X_1, F_1. Here if $X_1 = X$, then in order that the problem of solving (1.1.3) be well-posed it is necessary that F_1 be a subspace of F, whereas for $F_1 = F$ it is necessary that X_1 be an extension of X.

As for the questions related to ill-posed problems, there was a hypothesis that in problems of the type of the Cauchy problem for the Laplace equation it suffices to choose an appropriate pair of function spaces for the problem for this pair to be well-posed. Indeed, it is comparatively easy to choose the pairs of spaces in the Cauchy problem for the Laplace equation, in the Cauchy problem for the heat equation with reverse time, and also in many other problems of similar type so that the problems become well-posed.

However, this approach to ill-posed problems leaves aside an aspect which is very important from the point of view of applications. The fact is that if an equation of the type (1.1.3) is considered in connection with the mathematical modelling of a real physical phenomenon, then the right-hand side of the equation is often obtained on the basis of readings of physical instruments. Since the measurements are not exact, in such cases we cannot assume that the right-hand side of (1.1.3) is given with absolute accuracy. We may suppose only that we are given an element f_ε satisfying the inequality

$$\|f - f_\varepsilon\| \leq \varepsilon,$$

where the number ε is determined by the precision of the instruments. Here the norm of the space in which we know an estimate of the error of the right-hand side cannot be arbitrary: it is dictated by the organization of the system of measurements. As a rule, this is either the norm in the space C (if we know the estimate of the maximal error of the measurements) or the norm in L_2 (if we know the root-mean square error).

Although it involves additional technical difficulties, the organization of the system of measurements is possible when the error is small together with its derivative (the norm in the space C^1 or $W_2^{(1)}$). The organization of measurements when the error is small together with its second derivative (the spaces C^2 and $W_2^{(2)}$) either cannot be accomplished or has considerable technical difficulties.

Let us consider the integral equations (1.1.1), (1.1.2). Evidently, if solutions of these equations exist, then the right-hand sides of these equations have derivatives of all orders. Moreover, the right-hand sides are analytic functions in a certain complex neighborhood of the real axis, which improves the limitation on the order of growth of the norms of the derivatives.

Let F be a normed function space such that the problem of solving (1.1.1) is well-posed for a pair of spaces consisting of any ordinary space of solutions, for example, C or L_2, and the space F of right-hand sides. From what has been said above it follows that derivatives of all orders must participate in the definition of the norm in F, which is unacceptable from the viewpoint of real systems of measurements.

Suppose we consider a classical problem of mathematical analysis — the problem of differentiation. If a function is given exactly, in the form of a composition of elementary functions or integrals depending on a parameter, then the derivative can be also found exactly by the classical rules of differentiation. However, if the function is obtained on the basis of experimental data in the form of a table or graph, the problem of finding derivatives becomes more complicated. As was already noted, the error may be considered small in the spaces C or L_2, and it is desirable to obtain the derivative with an error which is small in these spaces. In this case, the problem of differentiation is a problem which is ill-posed in the classical sense.

Principles for the formulation of ill-posed problems that are natural from the point of view of applications were first stated in the work of A. N. Tikhonov in 1943, which was related to the justification of the trial-and-error method in the interpretation of data of geophysical measurements. This idea attracted the attention of many researchers, especially geophysicists.

A lot of ill-posed problems are easily reduced to operator equations of the first kind. We consider a more general equation.

Suppose X and F are complete metric spaces and $A(\cdot)$ is a continuous operator (mapping) defined on X with range F.

The equation

$$Ax = f, \tag{1.1.4}$$

where $x \in X$ and $f \in F$, is defined as the operator equation of the first kind if the operator A is completely continuous and any ball in the space X is not a compact set.

We now formulate the notion of classical well-posedness (following Hadamard) for equation (1.1.4).

Definition 1.1.1. The problem of solving (1.1.4) is well-posed in the classical sense if

1) for every $f \in F$ there exists a solution $x \in X$ of equation (1.1.4);

2) the solution of (1.1.4) is unique;

3) the solution x depends continuously on the right-hand side f. This means that small variations of f (in the metric of the space F) correspond to small variations of the solution x (in X).

For the classical problems of mathematical physics and the corresponding operator equations, the conditions of well-posedness are established by the existence theorems, the uniqueness theorems, and by continuous dependence of solutions on the data.

For the well-posedness of the problem of solving equation (1.1.4) it is necessary and sufficient that there exists a continuous operator B defined on F with range X which is inverse to the operator A.

The problem of solving an operator equation of the first kind cannot be well-posed since the operator inverse to a completely continuous operator is not continuous (see Lavrent'ev, Romanov, and Shishatskii, 1986).

The methods of solving ill-posed problems were proposed in the works of A. N. Tikhonov, M. M. Lavrent'ev, V. K. Ivanov, and R. S. Phillips. More detailed presentations of these methods can be found in Lavrent'ev, Romanov, and Shishatskii (1986), Tikhonov and Arsenin (1977).

Further we shall show that the problems of computerized tomography are ill-posed. We shall propose the algorithms of solving the problems of integral geometry and tomography constructed on the basis of the methods of solving ill-posed problems.

1.2. PROBLEM OF INTEGRAL GEOMETRY

We shall consider the following problem. Let $u(x)$ be a sufficiently smooth function defined in n-dimensional space $x = (x_1, ..., x_n)$, and let $\{\mathrm{M}(\lambda)\}$ be a family of smooth manifolds in this space depending on a parameter $\lambda = (\lambda_1, \ldots, \lambda_k)$. Suppose that we know the integrals

$$\int_{\mathrm{M}(\lambda)} u(x)\,\mathrm{d}\sigma = v(\lambda) \tag{1.2.1}$$

where $\mathrm{d}\sigma$ is the measure element on $\mathrm{M}(\lambda)$. Given the function $v(\lambda)$, it is required to find the function $u(x)$. In this case, the following questions arise. The first question is the most important: Does the function $v(\lambda)$ uniquely determine the function $u(x)$? The next question is: How can we find $u(x)$ in terms of $v(\lambda)$? Here, of course, it is desirable to obtain an analytic formula expressing $u(x)$ in terms of $v(\lambda)$. We observe, by the way, that this is possible only in exceptional cases. Finally, the last question is related in a natural way to the existence theorem for the problem: What are the necessary and sufficient conditions for $v(\lambda)$ to belong to the class of functions that can be represented in the form (1.2.1)?

As is evident from (1.2.1), the problem of integral geometry is the problem of solving a special integral equation of the first kind. There are problems among the problems of integral geometry which are strongly unstable. In this connection, estimating the conditional stability of the problem becomes of great importance. On the other hand, the question of constructing a function $u(x)$ from the function $v(\lambda)$ in the general case requires the development of special computational algorithms based on the general theory of ill-posed problems. In connection with this, it is expedient, for example, to construct the regularizers for the problem of integral geometry which take into account the specific features of the problem.

Problems of integral geometry in the formulation (1.2.1) were first considered at the beginning of our century. The Radon transform became widely known in mathematics. It associates a function $u(x)$ with its integrals $v(\lambda)$ over all possible hyperplanes. Finding the inverse transformation is the problem (1.2.1) of integral geometry for the case where $\{\mathrm{M}(\lambda)\}$ is the family of all possible hyperplanes in n-dimensional space. The problem was first solved by Radon; later it was considered in various aspects by John, Khachaturov, Kostelyanets and Reshetnyak, Gel'fand and Graev, and by some other authors.

The problem of integral geometry on linear manifolds turned out to be closely connected with the theory of representations of Lie groups. This fact explains why this problem attracts considerable interest. Among other problems of integral geometry the greatest popularity was enjoyed by problems of integral geometry on surfaces of the second order (spheres and ellipsoids). In the case when the family $\{\mathrm{M}(\lambda)\}$ is a family of spheres of arbitrary radius with centers which belong to a set of points of a fixed hyperplane, the problem was considered by Courant, while for a family of spheres of fixed radius it was considered by John. Various formulations of the problem of integral geometry on ellipsoids were considered by John, Plaksin, Romanov, Uspenskii, and Klibanov.

In the case when $\mathrm{M}(\lambda)$ is a more complex geometric object, problems of integral geometry were studied relatively recently. The incentive to this study was the connection found between problems of integral geometry and multidimensional inverse problems for differential equations. Lavrent'ev and Romanov were the first who focused the attention on this connection. Subsequently, Romanov constructed the examples in which the investigation of inverse problems for equations of hyperbolic type was reduced to the study of problems of integral geometry. The manifolds arising here are naturally connected with the original differential equation. These are either sections of characteristic conoids or projections of the bicharacteristics onto the space orthogonal to the time axis. For equations with variable coefficients, these manifolds are rather complex geometric objects.

Romanov obtained the first results in integral geometry for a rather general family of curves $\mathrm{M}(\lambda)$ in two-dimensional space which is invariant with respect to the group of rotations. This result was then generalized to the case of families of curves and hypersurfaces in n-dimensional space which are invariant with respect to parallel translations of these objects along a fixed plane. The theory of problems of integral geometry was developed further in the works of Anikonov, Lavrent'ev and Bukhgeim, Romanov, Mukhometov, and other authors.

The study of the problem of integral geometry from the practical point of view is stimulated, first of all, by interest in its important special case, namely, the inverse kinematic problem. In this case, M(λ) is the set of geodesic curves (seismic rays); $u(x)$ is the function inverse to the speed of the probing signal propagation in the medium under study; $v(\lambda)$ is the arrival time of the refracted waves. The numerical algorithms used for solution of the problem often were unstable and unjustified. The formulas of inversion were, as a rule, not constructive because of the rigorous mathematical conditions. In practice, the solution does not converge even for a fraction of percent of the error in the initial data. The development of new nontrivial methods of regularization and their application is not effective yet. So, effective solution of the inverse kinematic problem, in fact, is not constructed. In this case, only the result of solving the inverse kinematic problem in the linearized formulation mentioned above is satisfactory. Mukhometov carried out the data processing in the profile Baikal – Pamir using the linearized formulation (see Mukhometov, 1975).

Using the linearized formulation of the inverse kinematic problem and applying the method of computational tomography, it is possible to avoid solving the problem of integral geometry in the plane. In this case the problem was considered as the problem of integral geometry on a surface of the second order, and the constructive form of inversion based on the inversion of the two-dimensional Radon transform was obtained. An effective numerical algorithm of solving this problem was developed by Bukhgeim, Zerkal, and Pickalov (1983), Zerkal (1988), and Lavrent'ev *et al.* (1995). These results are presented in the third chapter of the book.

1.3. THE RADON TRANSFORM

The transformation of function $f(x) = f(x_1, \dots, x_n)$ given in n-dimensional space defined by the formula

$$\breve{f}(\xi, p) = \int f(x)\delta(p - (\xi, x))\,\mathrm{d}x \tag{1.3.1}$$

is called the Radon transform.

We have used the symbol of delta function in this formula. We integrate along the hyperplane $(\xi, x) = p$, where ξ is the normal vector to the hyperplane; p is the distance from the origin to the hyperplane; and (ξ, x) is the scalar product of the vectors ξ and x. The hyperplane has dimension $n - 1$.

In the X-ray computerized tomography, the cases $n = 3$ and $n = 2$ are of major interest. In the case $n = 3$, we integrate along the plane; in the case $n = 2$, we integrate along the line.

The formulas of inversion of the Radon transform are known (see Gel'fand, Graev, and Vilenkin, 1966; Helgason, 1980). These are the formulas which express the function $f(x)$ in terms of its Radon transform. The inversion formulas differ substantially for the cases of odd and even dimensions. For the case of odd-dimensional spaces, the inversion formula has the following form (see Gel'fand, Graev, and Vilenkin, 1966):

$$f(x) = C_n \int_\Gamma \check{f}_p^{(n-1)}(\xi, (\xi, x))\omega(\xi). \tag{1.3.2}$$

Here C_n is a constant dependent on the dimension of the space:

$$C_n = (-1)^{n-1}/(2(2\pi)^{n-1});$$

$\check{f}_p^{(n-1)}$ is the derivative of order $(n-1)$ with respect to the variable p; $\omega(\xi) = \sum_{k=1}^{n}(-1)^{k-1}\xi_k \, \mathrm{d}\xi_1 \ldots \mathrm{d}\xi_{k-1} \, \mathrm{d}\xi_{k+1} \ldots \mathrm{d}\xi_n$; Γ is a surface in the space ξ containing the point $\xi = 0$. The most natural surface of such kind is a sphere.

Now we shall clarify the geometric meaning of formula (1.3.2) for the three-dimensional case, which is important for applications. In order to calculate the value of the function at a point, we must do the following:

— to fix a point ξ on the unit sphere;

— to calculate the second derivative of the Radon transform with respect to p at the point (ξ, p) (in calculations, we use the integrals over the planes which are orthogonal to the vector ξ and pass through the point x and the points close to x);

— to average the results obtained for each point ξ over all points of the unit sphere, i.e., we have to integrate.

For the case $n = 2k$, the inversion formula for the Radon transform has the form

$$f(x) = C_n \int_\Gamma \int_{-\infty}^{\infty} \check{f}(\xi, (\xi, x))(p - (\xi, x))^{-n} \, \mathrm{d}p \, \omega(\xi). \tag{1.3.3}$$

Here Γ is a surface in the space ξ containing the point $\xi = 0$. Integration with respect to p in (1.3.3) is performed in the sense of generalized functions.

The classical computerized tomography corresponds to the case of $n = 2$. A three-dimensional body is represented as a collection of thin sections. Each section is a function of two variables and is reconstructed from the integral values of the density along the lines lying in the plane of the section.

In the more general case of three-dimensional tomography we consider the problem of reconstructing a function of three variables if we know its ray transformation — the integrals along a certain collection of rays in three-dimensional space. The case of the classical computerized tomography is the case when this collection consists of the rays lying in the plane $z = 0$ and the rays parallel to this plane. As a rule, in three-dimensional tomography, we consider the set of rays intersecting a given spatial curve which is the trajectory of the radiation source.

For the two-dimensional case, the Radon transform coincides with the ray transformation. In the three-dimensional case, the Radon transform is the integral along the planes, not along the lines.

The exact definition of the ray transformation for the three-dimensional space and the inversion formulas which allow constructing the algorithms will be given in the sections below.

It should be noted that some of the problems of geophysics are reduced specifically to the integration over the planes, i.e., to the inversion of the Radon transform in the three-dimensional space (see Alekseev and Tsibul'chik, 1985). Besides, some methods of inversion of the ray transformation are based on reducing the problem to the inversion of the Radon transform (see Three Dimensional Image Reconstruction in Radiology and Nuclear Medicine, 1995).

Further, we shall consider the numerical algorithms of inversion of the Radon transform and the ray transformation in two-dimensional and three-dimensional spaces.

1.4. RADON PROBLEM AS AN EXAMPLE OF AN ILL-POSED PROBLEM

We have considered the methods of numerical solution of ill-posed problems and presented the formulas of inversion of the Radon transform. We shall now consider the case of the Radon problem for a finite function of two variables, which is important in practice.

We suppose that $u(x, y)$ is a continuous function equal to zero outside the unit disk

$$x^2 + y^2 \leq 1. \tag{1.4.1}$$

We need to find the function $u(\cdot)$ knowing the integrals of this function along all the straight lines that intersect D.

We shall point out two representations of the Radon problem in the form of the problem of solving linear operator equations corresponding to the classical formulas of the parametrization of the systems of lines in the plane:

$$\int_{-\infty}^{\infty} u(x + s\cos\alpha, y + s\sin\alpha)\, \mathrm{d}s = f(x, y, \alpha) \tag{1.4.2}$$

$$\int_{-\infty}^{\infty}\int_{-\infty}^{\infty} u(\xi, \eta)\delta(x_0\xi + y_0\eta - \rho)\, \mathrm{d}\xi\, \mathrm{d}\eta = \varphi(x_0, y_0, \rho). \tag{1.4.3}$$

Since the integrals of $u(\cdot)$ along the lines which do not intersect D are equal to zero, we may consider the functions $f(\cdot)$ and $u(\cdot)$ to be given for all values of (x, y, α) and (x_0, y_0, ρ).

In the right-hand sides of equations (1.4.2), (1.4.3) we see the functions of three variables; however, they are determined in terms of functions of two variables.

The function $f(\cdot)$ satisfies the differential equation

$$f'_x \cos\alpha + f'_y \sin\alpha = 0.$$

Hence

$$f(x, y, \alpha) = f_0(x\sin\alpha - y\cos\alpha, \alpha).$$

The function $\varphi(\cdot)$, by the properties of the delta function, satisfies the relation

$$\varphi(x_0, y_0, \rho) = \frac{1}{r}\varphi\Big(\frac{x_0}{r}, \frac{y_0}{r}, \frac{\rho}{r}\Big) = \frac{1}{r}\psi\Big(\beta, \frac{\rho}{r}\Big) = \frac{1}{r}\psi(\beta, \tau),$$

where

$$\tau = \frac{\rho}{r}, \qquad x_0 = r\cos\beta, \qquad y_0 = r\sin\beta.$$

Representation (1.4.2) is the most natural from the practical point of view; representation (1.4.3) gives the simplest inversion formula. Evidently, we may pass from representation (1.4.2) to (1.4.3) and conversely using a change of variables.

By the property of $\varphi(\cdot)$ mentioned above, equation (1.4.3) is equivalent to the equation

$$\iint u(\xi, \eta)\delta(\xi\cos\beta + \eta\sin\beta - \tau)\, \mathrm{d}\xi\, \mathrm{d}\eta = \psi(\beta, \tau). \tag{1.4.4}$$

Because the function $u(\cdot)$ vanishes outside the disk D, we have

$$\psi(\beta, \tau) = 0, \quad |\tau| > 1.$$

Thus, the operator in (1.4.4) transforms the function $u(\cdot)$ given in the disk D into the function $\psi(\cdot)$ given in the rectangle $|\beta| \leq \pi$, $|\tau| \leq 1$. The function $\psi(\cdot)$ is the integral of the function $u(\cdot)$ along the line

$$\xi \cos\beta + \eta \sin\beta - \tau = 0$$

with the differential equal to the length element of this line.

Below we shall give some properties of the solution of equation (1.4.4).

1. The problem of solving (1.4.4) is ill-posed if $u(\cdot)$ and $\psi(\cdot)$ are the functions of the spaces L_2 or C.

We suppose that

$$u(x,y) = \begin{cases} 1, & x^2 + y^2 \leq \delta^2, \quad 0 < \delta < 1, \\ 0, & x^2 + y^2 > \delta^2. \end{cases}$$

Then

$$\psi(\beta,\tau) = \begin{cases} 2\sqrt{\delta^2 - \tau^2}, & |\tau| \leq \delta, \\ 0, \quad |\tau| > \delta, \end{cases}$$

$$\|u(\cdot)\|_{L_2} = \pi\delta^2, \qquad \|\psi(\cdot)\|_{L_2} = 8\,\frac{\pi}{\sqrt{3}}\,\delta^{3/2},$$

$$\|u(\cdot)\|_C = 1, \qquad \|\psi(\cdot)\|_C = 2\delta.$$

Thus, for δ sufficiently small, the relation of the norms of $u(\cdot)$ and $\psi(\cdot)$ will be arbitrarily large. This means that finite variation of $u(\cdot)$ may correspond to arbitrarily small variation of $\psi(\cdot)$.

The examples of ill-posed problems presented in previous sections lacked continuous dependence of solutions on the data in the case where the solutions and the data are considered as elements of the function spaces whose norms involve finite number of derivatives.

However, we may also call the problems ill-posed if for a certain pair of spaces there is no continuous dependence of the solution on the data, and for another pair of spaces there is a continuous dependence. In the theory of ill-posed problems, the problems for which there is no continuous dependence of the solution on the data for any pair of spaces with norms involving finite number of derivatives are called *strongly ill-posed*.

The problems in which there exist both kinds of pairs of spaces are called *weakly ill-posed*.

2. The problem of solving equation (1.4.4) is weakly ill-posed. Using the Fourier transform, we obtain

$$\iint e^{i(x_0\xi+y_0\eta)} u(\xi,\eta)\,d\xi\,d\eta = \int e^{ip}\varphi(x_0,y_0,p)\,dp$$

$$= \frac{1}{r}\int e^{ip}\psi\Big(\beta,\frac{p}{r}\Big)\,dp = \int e^{i\tau r}\psi(\beta,\tau)\,d\tau = \nu(\beta,r) \qquad (1.4.5)$$

where $\nu(\cdot)$ is the Fourier transform of the function $u(\cdot)$ in the polar coordinate system.

Note that the function $\nu(\cdot)$ in (1.4.5) is defined only for $r \geq 0$. However, the property of symmetry

$$\psi(-\beta,-\tau) = \psi(\beta,\tau)$$

allows us to define the function $\nu(\cdot)$ for all r by the formula

$$\nu(-\beta,-\tau) = \nu(\beta,\tau).$$

Parseval's equality for the functions $u(\cdot)$ and $\nu(\cdot)$ yields

$$\iint u^2(\xi,\eta)d\xi d\eta = \iint |\nu(\beta,r)|^2 r\,dr\,d\beta.$$

We denote by $H_{0\,1/2}$ the space of functions $\psi(\cdot)$ with the norm

$$\|\psi(\beta,r)\| = \iint |\nu(\beta,r)|^2 r\,dr\,d\beta,$$

where the function $\nu(\cdot)$ (the Fourier transform of the function $\psi(\cdot)$ with respect to the variable τ) is determined from (1.4.5).

Then we have

$$\|u(\cdot)\|_{L_2} = \|\psi(\cdot)\|_{H}.$$

The norm of the space $H_{0\,1/2}$ is defined in terms of the derivative of the function $\psi(\cdot)$ with respect to the variable τ to the power of $1/2$.

Here we also present a method of studying the question of uniqueness and stability for the Radon problem. Consider the problem of solving equation (1.4.2). We suppose that

$$F(x,y) = \int_{-\pi}^{\pi} f(x,y,\alpha)\,d\alpha.$$

The functions $F(\cdot)$ and $u(\cdot)$ are related by the formula

$$F(x,y) = \iint\limits_D \frac{1}{r}\, u(\xi,\eta)\, \mathrm{d}\xi\, \mathrm{d}\eta, \tag{1.4.6}$$

where $r = \sqrt{(x-\xi)^2 + (y-\eta)^2}$.

This equation may be considered as an integral equation of the first kind with weak singularity for the function $u(\cdot)$.

Now we obtain the inversion formula for equation (1.4.6). This formula actually uses the Laplace operator to a fractional power. We consider the function

$$w(x,y,x) = \iint\limits_D \frac{1}{R}\, u(\xi,\eta)\, \mathrm{d}\xi\, \mathrm{d}\eta,$$

where $R = \sqrt{(x-\xi)^2 + (y-\eta)^2 + z^2}$.

The function $w(\cdot)$ is the simple-layer potential with density concentrated in the disk D.

The well-known formulas of the potential theory yield

$$w(x,y,x) = \frac{1}{2\pi} \iint \frac{z}{R^3}\, F(\xi,\eta)\, \mathrm{d}\xi\, \mathrm{d}\eta,$$

$$u(x,y) = \frac{1}{2\pi} \left. \frac{\partial w(x,y,z)}{\partial z} \right|_{z=0}.$$

We now consider the Radon problem with incomplete data.

From the viewpoint of applications, the case of inverting the Radon transform is of interest when the integrals of the desired function are known not for all lines intersecting the domain.

Thus, we consider the problem of solving the operator equation (1.4.2) in the case when the right-hand side $f(\cdot)$ is given not for all variables (x,y) and α from the interval $|\alpha| \le \alpha_0 < \pi$.

In this case, we can show that the problem of solving (1.4.2) is strongly ill-posed.

We shall show here that the Radon problem with incomplete data can be reduced to the Cauchy problem for the Laplace equation.

If we reduce the solution of the Radon problem with incomplete data to equation (1.4.6), then the right-hand side $F(x,y)$ is known not for all values (x,y), but only for those which belong to a domain D_1 such that

$$D_1 \cap D = \emptyset.$$

In the space (x, y, z), the function $w(\cdot)$ considered outside the disk D will solve the Laplace equation. Evidently, we have

$$w'_z(x, y, 0) = 0, \qquad (x, y) \in D_1.$$

Thus, the Radon problem with incomplete data is reduced to determining the function

$$w(x, y, 0) = F(x, y), \qquad (x, y) \in D_1,$$
$$w'_z(x, y, 0) = 0,$$

i.e., to the Cauchy problem for the Laplace equation.

Remark. We have used the terminology of ill-posed problems proposed by Lavrent'ev. Natterer (1986) used a somewhat another classification. In essence, weakly ill-posed problems in the sense of Lavrent'ev are subdivided into weakly ill-posed and moderately ill-posed in Natterer's terminology.

The Radon problem for a function of two variables with finite support are moderately ill-posed in the sense of Natterer. The notions of strongly ill-posed problems coincide in both terminologies.

Figures 1.1 and 1.2 illustrate the statements formulated above, which provide mathematical justification of the possibility of constructing effective algorithms of computerized tomography. In Figure 1.1 the systems of disk holes and the corresponding Radon transforms are presented. In Figure 1.2 we have similar systems of holes and the squares of their Radon transforms. The systems of holes are chosen so that the norm in L_2 remains constant. We see that the norm in L_2 of the corresponding Radon transforms decreases, but not so rapidly.

1.5. THE ALGORITHM OF INVERSION OF THE TWO-DIMENSIONAL RADON TRANSFORM BASED ON THE CONVOLUTION WITH THE GENERALIZED FUNCTION $1/z^2$

In computerized X-ray tomography, a three-dimensional body is usually represented as a set of thin sections. To reconstruct the density of a section, we solve the problem of inverting the two-dimensional Radon transform. The Radon transform of the function $f(x, y)$ is defined to be the function

$$\check{f}(\xi_1, \xi_2, p) = \int f(x, y)\delta(p - x\xi_1 - y\xi_2)\, dx\, dy.$$

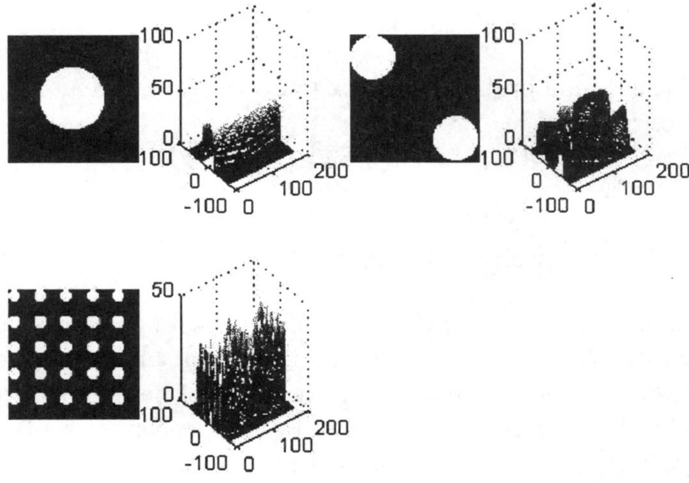

Figure 1.1

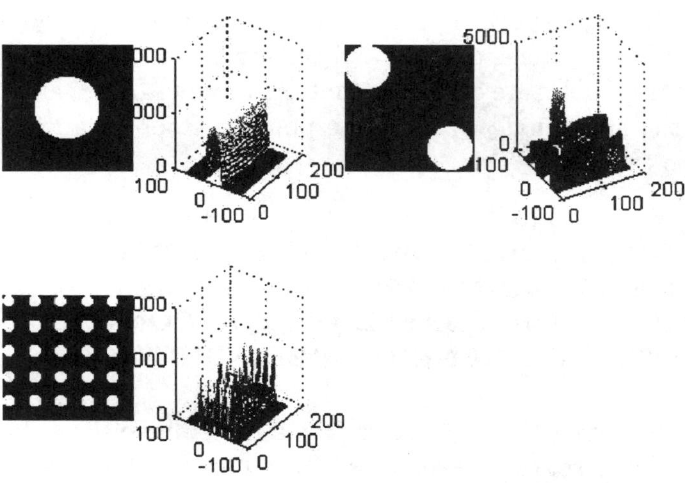

Figure 1.2

To reconstruct the function of two variables from the integrals of this function along the lines, we usually apply the method of convolution and inverse projection. In this method, the formula of inverting the Radon transform is written without using generalized functions. However, the inversion formulas have the most general and natural form when we use generalized functions. At the end of this section we shall consider the relation between the method of generalized functions and the method of convolution and inverse projection.

Before we describe the numerical algorithm, we shall derive the inversion formula which will show how to construct the algorithm.

Taking into account the equality

$$\check{f}(\lambda\xi, p) = \check{f}(\xi, p)/|\lambda|, \qquad (\xi = (\xi_1, \xi_2), \quad \lambda \neq 0)$$

we see that $\check{f}(\xi, p)$, for p fixed, is determined by the values ξ such that $|\xi| = 1$. This allows us to introduce the function

$$I(t, \varphi) = \check{f}(\cos\varphi, \sin\varphi, r) = \int_{L(r,\varphi)} f(x, y)\, \mathrm{d}l.$$

Here $L(r, \varphi)$ is the line orthogonal to the ray which forms the angle φ with the positive direction of the x-axis. The distance from the origin is r $(r \geq 0)$; if $r < 0$ then $L(r, \varphi)$ is the line symmetric to the line $L(|r|, \varphi)$ about the origin. We now express $f(x, y)$ in terms of $I(r, \varphi)$.

Since

$$f(x, y) = \frac{1}{4\pi^2} \iint \tilde{f}(\lambda_1, \lambda_2) \exp(-\mathrm{i}(x\lambda_1 + y\lambda_2))\, \mathrm{d}\lambda_1\, \mathrm{d}\lambda_2,$$

where $\tilde{f}(\lambda_1, \lambda_2)$ is the Fourier transform of the function f, then, passing to the polar coordinates and doing transformations with respect to φ in the interval $[\pi, 2\pi]$, we obtain

$$f(x, y) = \frac{1}{4\pi^2} \int_0^{\pi} \int_{-\infty}^{\infty} \tilde{f}(r\cos\varphi, r\sin\varphi) \exp(-\mathrm{i}r(x\cos\varphi + y\sin\varphi))|r|\, \mathrm{d}r\, \mathrm{d}\varphi.$$

Now, we introduce the function

$$S(z, \varphi) = \frac{1}{2\pi} \int_{-\infty}^{\infty} \tilde{f}(r\cos\varphi, r\sin\varphi) \exp(-\mathrm{i}rz)|r|\, \mathrm{d}r.$$

For a fixed φ, the function $S(z, \varphi)$ is the inverse one-dimensional Fourier transform of the product of $\tilde{f}(r\cos\varphi, r\sin\varphi)$ and $|r|$. For $\tilde{f}(r\cos\varphi, r\sin\varphi)$

the following equality holds (see Gel'fand, Graev, and Vilenkin, 1966):

$$\tilde{f}(r\cos\varphi, r\sin\varphi) = \int_{-\infty}^{\infty} I(z,\varphi)\exp(irz)\,\mathrm{d}z.$$

The inverse Fourier transform of $|r|$ is the generalized function $-1/\pi z^2$. Passing from the Fourier transform to the convolution, we obtain $S(z,\varphi) = I(z,\varphi) * (-1/\pi z^2)$. Using the regularization of the function $1/z^2$ (see Gel'fand and Shilov, 1959) we obtain the expression

$$S(z,\varphi) = \frac{1}{\pi}\int_0^{\infty} \frac{I(z+x,\varphi) + I(z-x,\varphi) - 2I(z,\varphi)}{x^2}\,\mathrm{d}x. \tag{1.5.1}$$

Thus, for $f(x,y)$ we obtain the formula

$$f(x,y) = \frac{1}{2\pi}\int_0^{\pi} S(x\cos\varphi + y\sin\varphi, \varphi)\,\mathrm{d}\varphi, \tag{1.5.2}$$

which allows us to express the desired function in terms of the data observed.

The inversion formula based on the generalized function $1/z^2$ for the Radon transform in $\mathbb{R}^2$ was given in another form in Herman and Tuy (1987).

Before we come to the discrete case, we make some remarks about the correctness of the algorithms considered above in real situations. The generalized functions are the functionals acting in the space of infinitely differentiable and rapidly decreasing functions. However, to construct the approximations of initial data (which are given at discrete points), it is desirable to have less restrictive conditions for the smoothness of the functions we approximate. The convolution with generalized functions, in particular, with the function $1/x^2$, may be defined for the functions which are considerably less smooth. This is very important for the justification of the numerical algorithms obtained with the use of the apparatus of generalized functions.

Now we consider the discrete case. We assume that $f(x,y) = 0$ outside the disk of radius R centered at the origin. The values $I(r_i,\varphi_j)$ are the initial data; r_i are the reference points in the interval $[-R,R]$, $1 \le i \le M$ and φ_j are the reference points in the interval $[0,\pi]$, $1 \le j \le N$. If we construct now the approximation $I(r,\varphi)$ using the values $I(r,\varphi)$ at the points (r_i,φ_j) so that $S(z,\varphi)$ satisfies (1.5.1), then, using (1.5.1) and (1.5.2), we can construct the approximation of $f(x,y)$.

We shall assume further that the reference points in the axes r and φ are equidistant.

For every φ_j we define $\tilde{I}(r, \varphi_j)$ as follows:

1. The function $\tilde{I}(r, \varphi_j)$ has continuous derivative with respect to r.

2. At the grid points, the function coincides with the given values and its derivative at these points is equal to the corresponding given values. This means that the following equalities hold:

$$\tilde{I}(r_i, \varphi_j) = I(r_i, \varphi_j),$$
$$\tilde{I}'(r_i, \varphi_j) = (I(r_{i+1}, \varphi_j) - I(r_{i-1}, \varphi_j))/2h.$$

Here $h = 2R/(M-1)$, $I(r_0, \varphi_j) = I(r_{M+1}, \varphi_j) = 0$, $j = 1, \ldots, M$.

3. The function $\tilde{I}(r, \varphi_j)$ is a polynomial of degree 3 with respect to r in the interval $[r_i, r_{i+1}]$.

These conditions allow us to obtain the coefficients of the corresponding spline explicitly. By immediate computation we obtain

$$\tilde{I}(r, \varphi_j) = \sum_{i=1}^{M} I(r_i, \varphi_j) Q(r + R + (1-i)h),$$

where

$$Q(x) = \begin{cases} (3/(2h^3))x^3 - (5/(2h^2))x^2 + 1, & \text{if } 0 < x < h, \\ (-1/(2h^3))x^3 + (5/(2h^2))x^2 - (4/h)x + 2, & \text{if } h < x < 2h, \end{cases}$$

$Q(x) = Q(-x)$, $Q(x) = 0$ for $|x| > 2h$, $h = r_{i+1} - r_i$. The second derivative of the function $Q(x)$ has discontinuities. However, the modulus of the second derivative is integrable. Taking this into account, we can show that the convolution $S_0(z) = Q(x) * (-1/\pi z^2)$ is expressed by formula (1.5.1). By immediate computation we obtain

$$S_0(z) = (-1/(h^3\pi^2))[2(3z + 2h)(z + h)\ln|z + h| - 9z^2 \ln|z|$$
$$+2(3z - 2h)(z - h)\ln|z - h| - 0.5(3z + 4h)(z + 2h)\ln|z + 2h|$$
$$-0.5(3z - 4h)(z - 2h)\ln|z - 2h|].$$

The graphs of the functions $Q(x)$ and $S_0(z)$ for various values of h are represented in Figures 1.3 and 1.4.

Thus, we have

$$\tilde{S}(z, \varphi) = \tilde{I}(z, \varphi) * (-1/z^2) = \sum_{i=1}^{M} I(r_i, \varphi_j) S_0(r + R + (1-i)h).$$

Substituting $\tilde{S}$ for S and a partial sum for the integral in (1.5.2), we obtain $f^*(x,y)$, the approximation of the function $f(x,y)$:

$$f^*(x,y) = \frac{1}{2\pi(M-1)} \sum_{i=1}^{M} \tilde{S}(x\cos\varphi_i + y\sin\varphi_i, \varphi_i). \tag{1.5.3}$$

As was noted above, in computerized tomography we usually consider the method of convolution and inverse projection. We now consider the relation between this method and the method presented in this section. Using integration by parts, the convolution with the generalized function $1/z^2$ can be replaced by differentiation and the convolution with the function $1/z$ (the Hilbert transformation). This means that the function

$$S(z,\varphi) = I(z,\varphi) * 1/z^2$$

can be represented as follows:

$$S(z,\varphi) = I'_z(z,\varphi) * 1/z.$$

When we construct the numerical algorithms, we use a certain sequence of regular functions $p_A(z)$ converging to $1/z$ (in the sense of generalized functions) as A tends to infinity. Using integration by parts, we differentiate the function $p_A(z)$. Thus, we obtain regular functions which converge to $1/z^2$. This means that the convolution with the generalized function $1/z^2$ is replaced by the sequence of convolutions with the regular functions $p'_A(z)$.

Thus, the stage of convolution in the classical method may be interpreted as follows: the initial data are approximated by a step function and the convolution is performed with a regular function which is an approximation of the generalized function $1/z^2$.

In the method presented in this section, the initial data are approximated by smooth functions, namely, by splines of the third order. This allows us to exactly calculate the convolution with the generalized function.

The stage of inverse projection corresponding to integration of the convolution is similar in both algorithms.

When we use the algorithm in real situations, it is important to estimate the influence of the noise on the accuracy of the approximation obtained. The explicit expression for the function we approximate allows us to calculate the dispersion of the estimate at each point for δr and $\delta\varphi$ fixed and for statistical characteristics of the noise that are known. For the case of independent additive stationary noise $\xi(z)$ we note the following. We consider

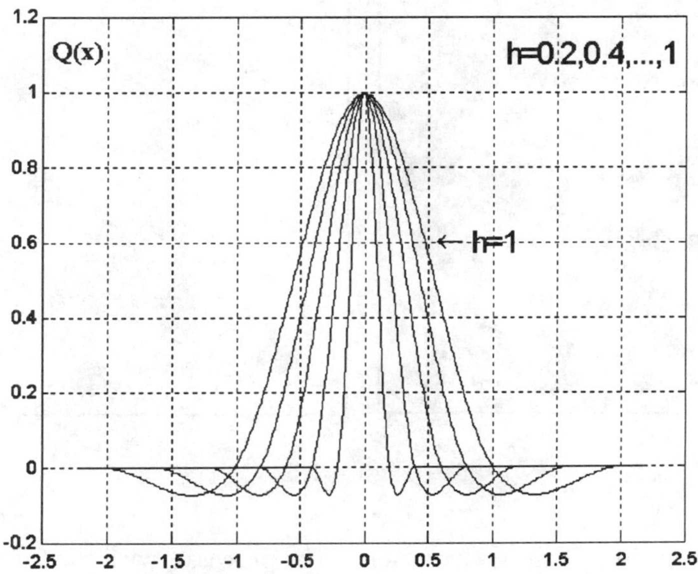

Figure 1.3

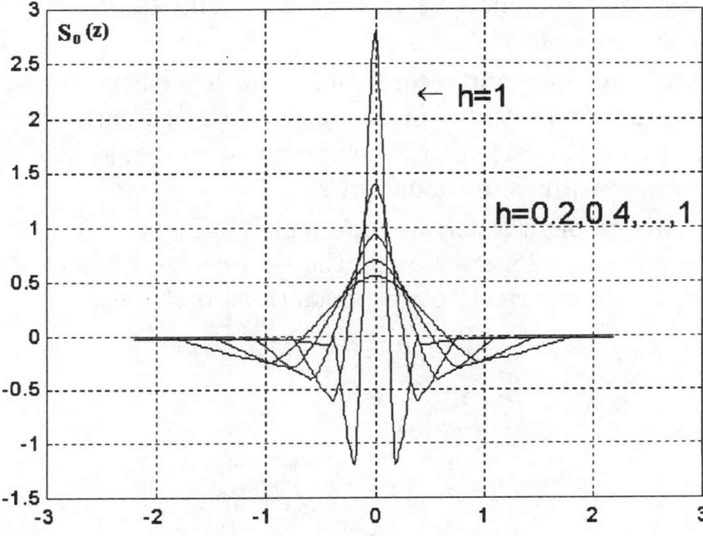

Figure 1.4

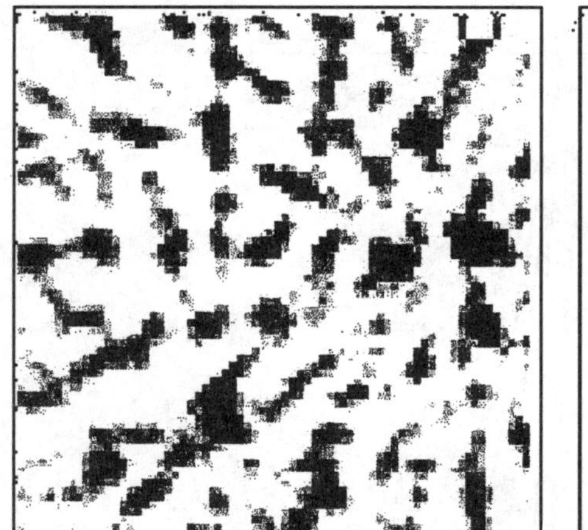
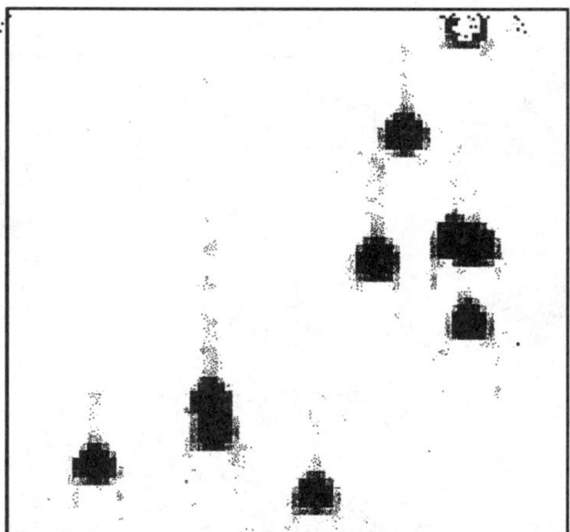

Figure 1.5

the process η which is the convolution of the process ξ with the function $1/z^2$. The spectral density of this linear transformation is $|\lambda|$. For the spectral densities of the processes ξ and η we have the relation $f_\eta(\lambda) = |\lambda|^2 f_\xi(\lambda)$. The dispersion of the process η is finite if $f_\eta(\lambda)$ is integrable. This means that the process ξ is mean square differentiable. In order that the convolution be expressed by formula (1.5.1), we should impose additional conditions. For example, the selected functions should have finite second derivative with probability equal to unity.

Numerical modelling and reconstruction of the density of real objects by applying the method presented in this section have demonstrated high accuracy of the method, in particular, when we investigate objects of complicated form and regions with sharp boundaries.

Examples of reconstruction are represented in Figure 1.5. The test object contains 10 particles. Figure 1.5 (on the left) corresponds to 10 rotations, and Figure 1.5 (on the right) corresponds to 32 rotations.

Chapter 2.

Cone-beam tomography reconstruction

2.1. REDUCING THE INVERSION FORMULAS OF CONE-BEAM TOMOGRAPHY RECONSTRUCTION TO THE FORM CONVENIENT FOR CONSTRUCTING NUMERICAL ALGORITHMS

A three-dimensional object in X-ray tomography is usually represented as a collection of thin sections. In order to reconstruct the density of the section, we solve the problem of inversion of the two-dimensional Radon transform. If we study a collection of objects, the following scheme is more natural: the radiation source moves along a certain spatial curve. Every point of the curve corresponds to a ray cone passing through this point. Initial data is the data on the attenuation of the radiation passing through the object. From the mathematical point of view, the problem is to reconstruct the function of three variables from the integrals along the curves passing through the point. The solution of this problem for various classes of functions and for various types of curves can be found in Anikonov (1978), Gel'fand and Goncharov (1986), Denisyuk (1991), Kirillov (1961), Orlov (1975), and Tuy (1983). The inversion formula for functions with finite support and for curves of special type was obtained by Tuy (1983). The main condition in this case is that the plane intersecting the object intersects the curve which is the trajectory of the source. The union of two circles lying in two orthogonal planes is an example of the curve satisfying the conditions of

Tuy (1983). However, the construction of numerical algorithms based directly on the formula of Tuy (1983) is difficult (see Natterer, 1986). One of the reasons is that the inversion formula is based on the Fourier transform of the homogeneous function obtained from the initial data. In this case the Fourier transform is understood in the sense of generalized functions, while the Fourier transform in the classical sense may not exist. In this section we shall present an expression for the Fourier transform which allows us to use the method stated in the previous section for constructing numerical algorithms. We also show the connection between the results of Kirillov (1961) and Tuy (1983).

We suppose that the function $f(x) = f(x_1, x_2, x_3)$, the point $S = (s_1, s_2, s_3)$ and the vector $\alpha = (\alpha_1, \alpha_2, \alpha_3)$ are given. The ray transformation of the function $f(x)$ is given by the function

$$(R_1^+ f)(\alpha, S) = \int_0^\infty f(t\alpha + S)\,\mathrm{d}t,$$

which is the integral of $f(x)$ along the ray passing through the point S along the vector α. In certain situations, along with the function $(R_1^+ f)(\alpha, S)$ we shall consider the function

$$(R_1 f)(\alpha, S) = \int_{-\infty}^\infty f(t\alpha + S)\,\mathrm{d}t$$

which is the integral along the entire line or, equivalently, the sum of integrals along the rays coming from the point S in the directions α and $-\alpha$.

The set of points S for which we know the ray transformation is usually the set of points belonging to a certain curve being the trajectory of the source of radiation.

We suppose that the curve $\Phi(\lambda) = (\Phi_1(\lambda), \Phi_2(\lambda), \Phi_3(\lambda))$ along which the source is moving is given. Here the parameter λ varies within a certain interval Λ of the real axis. For any $\alpha = (\alpha_1, \alpha_2, \alpha_3)$ and $\lambda \in \Lambda$, we define the function

$$g^+(\alpha, \lambda) = (R_1^+ f)(\alpha, \Phi(\lambda)) = \int_0^\infty f(t\alpha + \Phi(\lambda))\,\mathrm{d}t.$$

The function $g^+(\alpha, \lambda)$ is the integral of the function $f(x)$ along the ray with vertex $\Phi(\lambda)$ and direction vector α. Note that for any fixed λ the function $g^+(\alpha, \lambda)$ is a λ-homogeneous function of α of degree (-1):

$$g^+(\alpha, \lambda) = \frac{1}{|\alpha|} \int_0^\infty f\Big(\Phi(\lambda) + \frac{\alpha}{|\alpha|}\Big)\,\mathrm{d}t. \qquad (2.1.1)$$

For a function with finite support, in Tuy (1983) the following formula is obtained

$$f(x) = \int_0^{2\pi} \int_{-\pi/2}^{\pi/2} \cos\phi \frac{1}{2i\pi(\Phi'(\lambda), \beta)} \frac{\partial G^+(\beta, \lambda)}{\partial \lambda} \, d\phi \, d\theta. \tag{2.1.2}$$

For λ fixed, the function $G^+(\beta, \lambda)$ is the Fourier transform of the function $g^+(\alpha, \lambda)$ with respect to α; $\beta = (\cos\theta\cos\phi, \sin\theta\cos\phi, \sin\phi)$. In formula (2.1.1), λ depends on x and β and is chosen on the basis of the following conditions: the scalar product (β, x) is equal to $(\beta, \Phi(\lambda))$, but $(\beta, \Phi(\lambda))$ is not equal to zero. The value of the function $f(x)$ can be reconstructed at the point x, if such a λ exists for every β. Geometrically, this means that every plane which intersects the point x of the support of the function intersects the curve $\Phi(\lambda)$ so that the denominator in (2.1.2) is not equal to zero. The example of the curve satisfying the Kirillov – Tuy condition is the union of two circles lying in two orthogonal planes, provided that the support lies in the unit ball. A helix can be used for cylindrical objects.

The Fourier transform $G^+(\beta, \lambda)$ of the function $g^+(\alpha, \lambda)$ enters formula (2.1.2); however, the Fourier transform

$$G^+(\xi, \lambda) = \int_{-\infty}^{\infty} g^+(\alpha, \lambda) \exp(2i\pi(\alpha, \xi)) \, d\alpha$$

in the classical sense does not exist since $g^+(\alpha, \lambda)$ is homogeneous and has order $1/|\alpha|$ at infinity. The Fourier transform here is understood in the sense of generalized functions. Because $g^+(\alpha, \lambda)$ is a homogeneous function, for any fixed λ, the initial data are defined completely by their values on the surface $|\alpha| = 1$. If we use the Fourier transform, then we come to generalized functions when passing to the function in the whole space $\mathbb{R}^3$. The Fourier transform in the sense of generalized functions is a linear functional defined in a certain space. A more detailed presentation will be given in the following sections. It should be noted that not every functional is expressed by means of a regular function. In order to use the formulas of the form (2.1.2) for constructing algorithms, it is necessary to show that $G^+(\xi, \lambda)$ is expressed by means of a regular function and we must have the expression of this function in terms of the function $g^+(\alpha, \lambda)$. In Tuy (1983) an expression is given that relates $G^+(\xi, \lambda)$ to the desired function $f(x)$ for $\xi \neq 0$. This means that the functional $G^+(\xi, \lambda)$ is expressed by means of a regular function. However, in order to construct the algorithms of tomography reconstruction, we must express $G^+(\xi, \lambda)$ in terms of the initial data $g^+(\alpha, \lambda)$ and not in terms of the desired function $f(x)$.

So, we shall find $G^+(\xi, \lambda)$. We shall use the fact that $g^+(\alpha, \lambda)$ is a homogeneous function with respect to α for λ fixed. The following statement is proved in Trofimov (1993a).

Statement. *Suppose $G^+(\xi)$ is the Fourier transform in the sense of generalized functions of a homogeneous function $g^+(\alpha)$. Then*

$$G^+(\xi) = -\frac{1}{4\pi^2}\int_{|\beta|=1} g^+(\beta)[(\beta,\xi)^{-2} - \mathrm{i}\pi\delta'((\beta,\xi))]\,\mathrm{d}^2\beta. \qquad (2.1.3)$$

The rigorous proof involves the essential use of the apparatus of generalized functions understood in the sense of linear functionals over the corresponding space. Here we shall give the main ideas of the proof. In particular, we shall make a change of variables in divergent integrals similar to that in convergent integrals.

We represent $G^+(\xi)$ as follows:

$$G^+(\xi) = \int g^+(\alpha)\exp(-2\mathrm{i}\pi(\alpha,\xi))\,\mathrm{d}\alpha$$

(since the parameter λ is fixed, we can omit it).

As was mentioned above, the integral is divergent; nevertheless, passing to spherical coordinates as usual, we obtain

$$G^+(\xi) = \int_0^{2\pi}\int_{-\pi/2}^{\pi/2}, \cos\varphi\int_0^{\infty}\rho^2 g^+(\rho\beta)\exp(-2\mathrm{i}\pi\rho(\beta,\xi))\,\mathrm{d}\rho\,\mathrm{d}\varphi\,\mathrm{d}\theta$$

where $\beta = \beta(\varphi,\theta) = (\cos\theta\cos\varphi, \sin\theta\cos\varphi, \sin\varphi)$, $\varphi \in [-\pi/2, \pi/2]$, $\theta \in [0, \pi]$. Taking into account that $g^+(\rho\beta) = g^+(\beta)/\rho$ and the fact that integrating with respect to the angles φ and θ corresponds to integrating over the unit sphere, we obtain

$$G^+(\xi) = \int_{|\beta|=1} g^+(\beta)\left[\int_0^{\infty}\rho\exp(-2\mathrm{i}\pi\rho(\beta,\xi))\,\mathrm{d}\rho\right]\mathrm{d}^2\beta.$$

The integral with respect to ρ is the Fourier transform of ρ_+. Using the tables for Fourier transforms of generalized functions (see Gel'fand and Shilov, 1959) we obtain expression (2.1.3).

For real functions $f(x)$ in (2.1.2) we need the imaginary part of $G^+(\xi)$:

$$\operatorname{Im} G^+(\xi) = \frac{\mathrm{i}}{4\pi^2}\int_{|\beta|=1} g^+(\beta)\delta'(\beta,\xi)\,\mathrm{d}^2\beta.$$

Using the generalized functions concentrated on the surface (see Gel'fand and Shilov, 1959) we have

$$\operatorname{Im} G^{+}(\xi) = \frac{\mathrm{i}}{4\pi^2} \int_{S(\xi)} L(\xi, D) g^{+}(\gamma)\, \mathrm{d}\gamma.$$

Here $S(\xi) = \{\gamma \in S^2 \mid (\xi, \gamma) = 0\}$, $L(\xi, D) = \sum_{k=1}^{3} \xi_k \frac{\partial}{\partial \gamma_k}$ is the derivative in the direction ξ. Substituting the functions $g^{+}(\xi, \lambda)$ and $G^{+}(\xi, \lambda)$ dependent on the parameter λ into (2.1.2), we obtain the following inversion formula useful for constructing numerical algorithms:

$$f(x) = \frac{-1}{8\pi^2} \int_0^{2\pi} \int_{-\pi/2}^{\pi/2} \cos\theta \, \frac{1}{\langle \Phi'(\lambda), \beta \rangle} \frac{\partial}{\partial \lambda} \Big[\int_{S(\xi)} L(\beta, D) g^{+}(\lambda, \xi) \Omega(\xi) \Big] \, \mathrm{d}\theta \, \mathrm{d}\varphi. \tag{2.1.4}$$

Here $S(\xi)$ is the circle which is the intersection of the unit sphere and the plane $P(\beta)$. The plane $P(\beta)$ passes through the origin and is orthogonal to the vector β. The symbol $\Omega(\xi)$ means that integration is performed along the circle. The operator $L(\beta, D)$ means differentiation of the function $g^{+}(\xi, \lambda)$ in the direction of the vector β:

$$L(\beta, D) g^{+}(\xi, \lambda) = \beta_1 \frac{\partial}{\partial \xi_1} g^{+}(\xi, \lambda) + \beta_2 \frac{\partial}{\partial \xi_2} g^{+}(\xi, \lambda) + \beta_3 \frac{\partial}{\partial \xi_3} g^{+}(\xi, \lambda).$$

Here λ is fixed and depends on β and x.

As before, we have $\beta = \beta(\varphi, \theta) = (\cos\theta\cos\varphi, \sin\theta\cos\varphi, \sin\varphi)$ and $\lambda = \lambda(\theta, \varphi) = \lambda(x, \beta)$ is such that the scalar product (x, β) is equal to $(\beta, \gamma(\lambda))$ and $(\beta, \gamma'(\lambda)) \neq 0$.

In formula (2.1.4) we use regular functions, which is useful for constructing numerical algorithms.

Remark. Denisyuk (1991) obtained the inversion formulas for g^{+} in $\mathbb{R}^n$ independently and without explicit use of the Fourier transform of generalized functions. For $n = 3$, Denisyuk's formulas and the formulas obtained with the use of Tuy's formulas coincide.

Above, we have obtained the formulas which allow us to construct numerical algorithms of reconstruction of the function $f(x) = f(x_1, x_2, x_2)$ from its ray transformation

$$(R_1^{+} f)(\alpha, S) = \int_0^{\infty} f(t\alpha + S)\, \mathrm{d}t.$$

In what follows we shall omit the symbol f and use the notation

$$(R_1^+ f)(\alpha, S) = R_1^+(\alpha, S).$$

For S fixed, the function $R_1^+(\alpha, S)$ is a function in three-dimensional space; however, as it is a homogeneous function, there exist surfaces such that $R_1^+(\alpha, S)$ is completely determined by its values on these surfaces (surfaces on which the radiation receivers are situated).

It is convenient to consider the initial data in the form $R_1^+(\alpha, S)$ if the matrix of receivers is situated on a sphere. However, in real situations the matrix of receivers is usually situated in the plane or on the cylinder surface. In these cases it is more convenient to use another form of initial data.

The planar detector.

We suppose that for the source concentrated at the point $S = (s_1, s_2, s_3)$ the initial data are registered in the plane P defined by the equation $xs_1 + ys_2 + zs_3 = -|S|$. The plane P is defined by the following conditions:

— the plane P is orthogonal to the ray connecting the source with the origin;

— the plane P passes through the point $S = (s_1, s_2, s_3)$.

The distance D from the source to the plane of registration is equal to the double distance from the source to the origin. In the plane of registration, we shall use the Cartesian system (p_1, p_2) whose origin is at the point of intersection of the ray connecting the source and the point $(0, 0, 0)$. Thus, if the source is at the point $S = (s_1, s_2, s_3)$, then the origin of the coordinate system (p_1, p_2) in the plane of observation is situated at the point with the three-dimensional coordinates $(-s_1, -s_2, -s_3) = -S$.

In the case of the cone-beam reconstruction, the most widely known examples of the source trajectories are a helix and the union of two circles lying in intersecting planes.

The trajectory in the form of two circles.

We consider a circle lying in the plane $z = 0$.

The direction of the axis p_2 in the plane of registration is assumed to coincide with the direction of the axis z.

For the axis p_1 of the coordinate system we take the line of intersection of the plane of registration and the plane containing the circle along which the source is moving. In order to define the coordinate system completely, we need to choose one of the two possible directions of the axis p_1. If $s_3 = 0$, $s_1 = r\cos\lambda$, $s_2 = r\sin\lambda$ (the source is moving in the plane $z = 0$), then the positive unit vector in the axis p_1 is chosen so that it coincides with the vector $(\cos(\lambda + \pi/2), \sin(\lambda + \pi/2), 0) = (-\sin\lambda, \cos\lambda, 0) = (-s_2/|S|, s_1/|S|, 0)$.

The point whose coordinates in the plane of registration are (p_1, p_2) has the following spatial coordinates:

$$
\begin{aligned}
x &= -p_1 \sin\lambda - r\cos\lambda = \frac{-p_1 s_2}{|S|} - s_1,\\
y &= p_1 \cos\lambda - r\sin\lambda = \frac{p_1 s_1}{|S|} - s_2,\\
z &= p_2.
\end{aligned}
$$

In the case of the planar detector, the initial data are the integrals along the rays connecting the points (p_1, p_2) in the plane of registration with the source S.

The function $g_r(p_1, p_2, \lambda)$ being registered is the integral of the desired function $f(x) = f(x_1, x_2, x_3)$ along the ray coming from the point $S = (s_1, s_2, s_3) = (r\cos\lambda, r\sin\lambda, 0)$ in the direction of the point

$$
\begin{aligned}
P &= (-p_1 \sin\lambda - r\cos\lambda, p_1\cos\lambda - r\sin\lambda, p_2)\\
&= \left(\frac{-p_1 s_2}{|S|} - s_1, \frac{p_1 s_1}{|S|} - s_2, p_2\right).
\end{aligned}
$$

The integral form of this function is as follows:

$$
\begin{aligned}
g_r(p_1, p_2, \lambda) = \int_0^\infty & f(r\cos\lambda + t(-p_1\sin\lambda - 2r\cos\lambda),\\
& r\sin\lambda + t(p_1\cos\lambda - 2r\sin\lambda), 0 + tp_2)\, dt.
\end{aligned}
$$

For $t = 0$, the ray passes through the point $S = (r\cos\lambda, r\sin\lambda, 0)$; and for $t = 1$, the ray passes through the point $P = (p_1, p_2) = (-p_1\sin\lambda - r\cos\lambda, p_1\cos\lambda - r\sin\lambda, p_2)$.

Thus, we have obtained the relation between the functions $g_r(p_1, p_2, \lambda)$ and $R_1^+(\xi, S(\lambda))$:

$$
\begin{aligned}
g_r(p_1, p_2, \lambda) &= R_1^+(-p_1\sin\lambda - 2r\cos\lambda, p_1\cos\lambda - 2r\sin\lambda, p_2, S(\lambda))\\
S(\lambda) &= (s_1(\lambda), s_2(\lambda), s_3(\lambda)) = (r\cos\lambda, r\sin\lambda, 0).
\end{aligned}
$$

Along with the notation $g_r(p_1, p_2, \lambda)$ we shall use the designations $g_r(p_1, p_2, S(\lambda))$, $g_r(p_1, p_2, S)$, $g_r(P, S)$. Here $S(\lambda)$ is the point in the trajectory of the source corresponding to the parameter λ; $P = (p_1, p_2)$. We have expressed the function $g_r(p_1, p_2, \lambda)$ in terms of the function $R_1^+(\xi, S(\lambda)) = g^+(\xi, \lambda)$.

In the inversion formula for the ray transformation we have used the function $g^+(\xi, \lambda) = R_1^+(\xi, S(\lambda))$. In order to use $g_r(p_1, p_2, \lambda)$, which is registered in the case of the planar detector, we need to express $g^+(\xi, \lambda)$ in terms of $g_r(p_1, p_2, \lambda)$.

Further we shall need the coordinates (p_1, p_2) (in the coordinate system of the registration plane) of the point of intersection of the data registration plane and the ray $(S + t\xi) = (s_1 + t\xi_1, s_2 + t\xi_2, s_3 + t\xi_3)$. These coordinates are

$$\begin{aligned} p_1(S, \xi) &= \frac{1}{|S|}\Big[- s_2\Big(s_1 - \xi_1 \frac{2\,|S|^2}{\langle S, \xi \rangle}\Big) + s_1\Big(s_2 - \xi_2 \frac{2\,|S|^2}{\langle S, \xi \rangle}\Big)\Big] \\ &= \frac{2\,|S|\,(s_2\xi_1 - s_1\xi_2)}{\langle S, \xi \rangle} = \frac{2\,|S|\,(s_2\xi_1 - s_1\xi_2)}{s_1\xi_1 + s_2\xi_2}, \\ p_2(S, \xi) &= -\xi_3 \frac{2\,|S|^2}{\langle S, \xi \rangle}. \end{aligned}$$

Now we can express $R_1^+(\xi, S(\lambda))$ using $g_r(p_1, p_2, \lambda)$:

$$R_1^+(\xi, S(\lambda)) = \begin{cases} g^+(\xi, \lambda), & \text{if} \quad \langle S(\lambda), \xi \rangle < 0, \\ 0, & \text{if} \quad \langle S(\lambda), \xi \rangle \ge 0, \end{cases}$$

$$g^+(\xi, \lambda) = g_r\Big(\frac{2\,|S(\lambda)|\,(s_2\xi_1 - s_1\xi_2)}{\langle S(\lambda), \xi \rangle}, \frac{-2\,|S(\lambda)|^2\,\xi_3}{\langle S(\lambda), \xi \rangle}, \lambda\Big).$$

Thus, we have obtained the following relation between the functions $g^+(P, \lambda)$ and $R_1^+(\xi, S(\lambda)) = g^+(\xi, \lambda)$ ($P = (p_1, p_2)$; $\xi = (\xi_1, \xi_2, \xi_3)$):

$$R_1^+(\xi, S(\lambda)) = \begin{cases} g^+(\xi, \lambda), & \text{if} \quad \langle S(\lambda), \xi \rangle < 0, \\ 0, & \text{if} \quad \langle S(\lambda), \xi \rangle \ge 0, \end{cases}$$

$$g^+(\xi, \lambda) = g_r\Big(\frac{2\,|S(\lambda)|\,(s_2\xi_1 - s_1\xi_2)}{\langle S(\lambda), \xi \rangle}, \frac{-2\,|S(\lambda)|^2\,\xi_3}{\langle S(\lambda), \xi \rangle}, \lambda\Big).$$

If we pass from the function $g^+(\xi, \lambda) = R_1^+(\xi, S(\lambda))$ to the function $g_r(P, S)$, the integration along the circle $S(\lambda)$ in three-dimensional space is changed for the integration along straight lines in the plane of registration. Note that the formulas of inversion of the ray transformation, which involve integration along the lines in the plane of registration, are also given in Grangeat (1990).

In Figure 2.1 we present the results of computer simulation of reconstructing the ray transformation of a homogeneous sphere with a defect. The trajectory of the source consists of two circles lying in the planes $z = 0$ and $y = 0$. Four sections are represented. If we apply the classical approach, which reduces the problem to the two-dimensional one, in order to obtain four sections, the source trajectory must consist of four circles. It should be noted that the sphere corresponding to the defect is shifted relative to the centre of the main sphere. The domain is represented where the defect is supposed to be concentrated.

In Figure 2.2 the two sections from Figure 2.1 are presented. However, at first, we have processed the images using nonlinear (median) filters. This method is given in application to the tomography problems in more detailed version in Vainikko and Veretennikov (1986).

In Figures 2.3–2.5 the scheme of scanning is represented and the simulation results for the reconstruction of a layer of a cylindrical object (a piece of a pipe with defects in the walls) are given. The pipe is full. The density of the pipe walls is equal to 4; the density of the filling stuff is equal to 3; the density of the defect is equal to zero.

The trajectory of the source consisted of two circles in all these cases. This provided the fulfillment of the Kirillov – Tuy conditions for the part of object in question. The Kirillov – Tuy conditions are not necessary conditions for inverting the ray transformation. The trajectory can consist of only one circle (see Anikonov, 1978; Blagoveshchenskii, 1986; Finch, 1985). In this case, the problem in the whole will be unstable in the Sobolev spaces. However, the problem can be divided in two parts (see Blagoveshchenskii, 1986; Finch, 1985):

— the reconstruction of integrals of a function along the lines passing through the disk from the integrals along the lines passing through the circle;

— the reconstruction of the desired function of three variables.

It is shown in Trofimov (1992) that stable algorithms can be constructed for the first part of the problem. In some problems of flaw detection, it suffices to solve the first part of the problem, which is stable. In Figure 2.6 we see

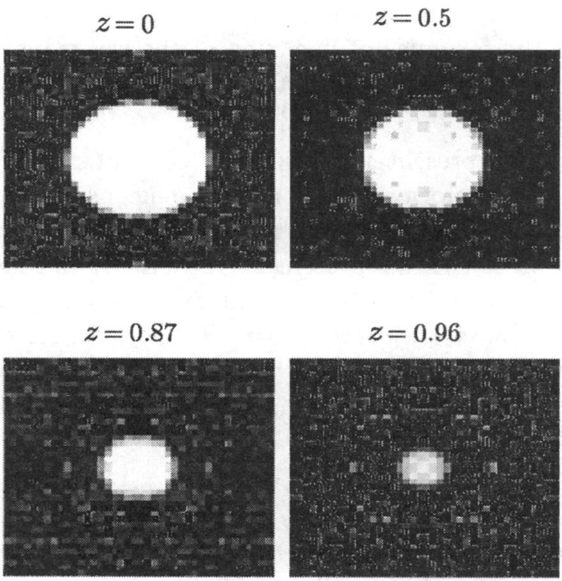

Figure 2.1: The results of computer simulation of reconstruction of the homogeneous sphere with a defect by the ray transformation. The object in the whole is not symmetric. The centre of the sphere corresponding to the defect in shifted relative to the centre of the main sphere. The domain where the defect is supposed to be situated is determined.

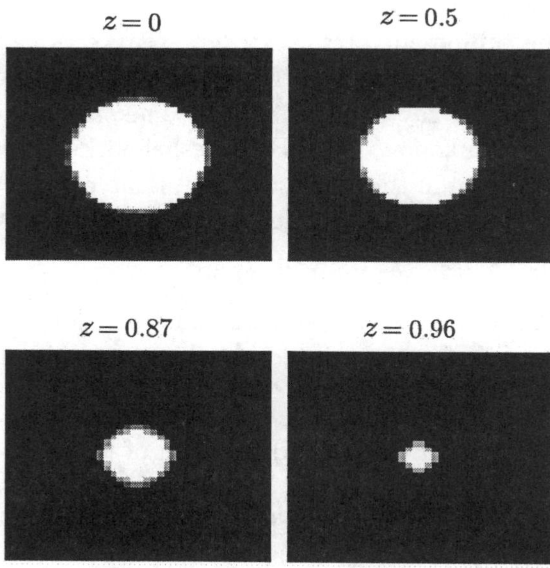

Figure 2.2: The sections of Figure 2.2 after processing images using nonlinear (median) filters.

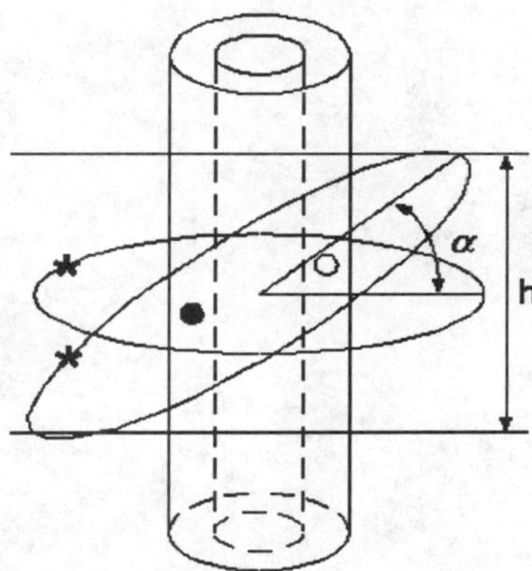

Figure 2.3: The scheme of scanning for reconstruction of a layer of cylindrical object (the segment of the pipe with defects in the walls).

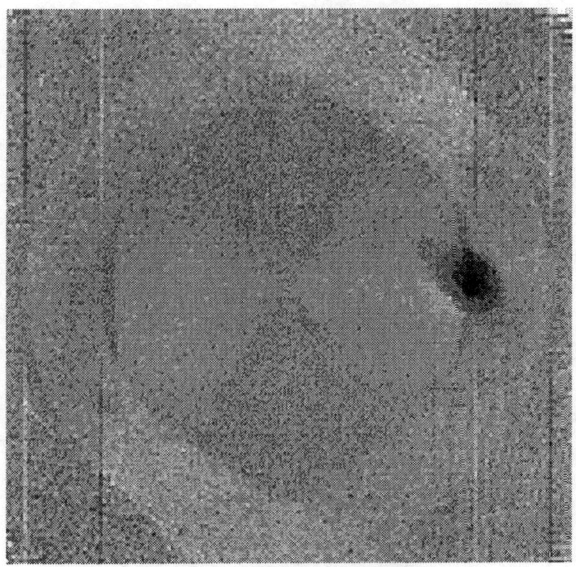

Figure 2.4: The result of simulation. The pipe is full, the density of the walls is equal to 4; the density of the filling stuff is equal to 3; the density of the defect is equal to 0.

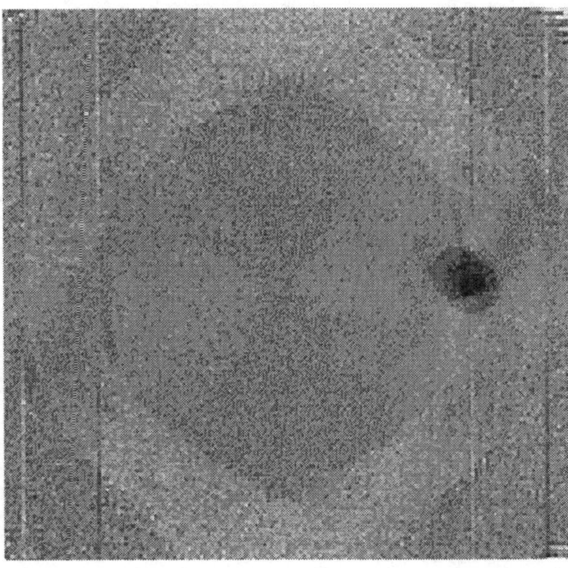

Figure 2.5: Intermediate result of reconstruction for the partial angle of rotation. Defects are seen already at this stage.

Figure 2.6: The example of reconstruction of integrals along the straight lines orthogonal to the plane $z = 0$ from the integrals along the lines passing through the unit circle in the plane $z = 0$. The defects in the walls are seen vividly (the pipe is empty in this case).

an example of simulation of reconstructing the integrals along the straight lines orthogonal to the plane $z = 0$ from the integrals along the straight lines passing through the unit circle in the plane $z = 0$. The defects in the wall are seen clearly (the pipe is empty in this case).

2.2. ELEMENTS OF THE THEORY OF GENERALIZED FUNCTIONS IN APPLICATION TO PROBLEMS OF INVERSION OF THE RAY TRANSFORMATION

A generalized function is a continuous linear functional defined in the space K of all functions $\alpha(x)$ that have derivatives of all orders and finite support (unique for each function $\alpha(x)$). Any regular integrable function $f(x)$ determines the linear functional (f, α):

$$(f, \alpha(x)) = \int_{-\infty}^{\infty} f(x)\alpha(x)\,\mathrm{d}x. \tag{2.2.1}$$

However, in the space K there exist continuous linear functionals which cannot be defined by means of regular integrable functions. The most well-known examples of such functions are the delta function and its derivatives. Another example is the functional based on the function $1/x$. The function $1/x$ is regular, but not integrable. If the corresponding functional is given, then the integral

$$\Big(\frac{1}{x}, \alpha(x)\Big) = \int_{-\infty}^{\infty} \frac{\alpha(x)}{x}\,\mathrm{d}x \tag{2.2.2}$$

is understood in the sense of principal value:

$$\Big(\frac{1}{x}, \alpha(x)\Big) = \int_{-\infty}^{\infty} \frac{\alpha(x)}{x}\,\mathrm{d}x = \lim_{\varepsilon\to 0}\int_{-\infty}^{-\varepsilon} \frac{\alpha(x)}{x}\,\mathrm{d}x + \lim_{\varepsilon\to 0}\int_{\varepsilon}^{\infty} \frac{\alpha(x)}{x}\,\mathrm{d}x.$$

This sense of integral is used for the definition of the Hilbert transformation of the function $\alpha(x)$ as of the convolution with the function $1/x$.

The Hilbert transformation is used, in particular, in one of the formulas of inversion of the Radon transform in two-dimensional space. This formula usually appears in the books devoted to X-ray tomography. However, the convolution and inverse projection method, which is often used when constructing numerical algorithms of tomography reconstruction, is based on a somewhat another form of the formula of inversion of the Radon transform. In this method, the convolution of projection data with a sequence of functions converging to $1/x^2$ in the sense of generalized functions is essentially used.

The linear functional corresponding to $1/x^2$ or, equivalently, the generalized function $1/x^2$, is defined by the following formula (see Gel'fand and Shilov, 1959):

$$\left(\frac{1}{x^2}, \alpha(x)\right) = \int_0^\infty \frac{\alpha(-x) + \alpha(x) - 2\alpha(0)}{x^2}\,\mathrm{d}x. \tag{2.2.3}$$

The integral in (2.2.3) is convergent in the ordinary sense for every function $\alpha(x)$ from the space of the basic functions. It is convergent even for a wider class of functions.

In the formulas of inversion of the Radon transform we use the convolution with the function $1/x^2$. The convolution of generalized functions is defined as follows.

We suppose that two functionals f and g are given. The action of the functional $f * g$, which is the convolution of f and g, on the function $\alpha(x)$ from the space of the basic functions is defined by the formula

$$(f * g, \alpha) = (f_x, (g_y, \alpha(x+y))). \tag{2.2.4}$$

Here g_y means that the functional acts on the function α as the function of y, and the functional f acts on the resulting function of x. If we specify the functionals f and g using regular functions, then the convolution functional defined by formula (2.2.4) may be specified using the function which is the convolution of the corresponding functions in the ordinary sense.

Here the following remark is necessary. If a function $\alpha(\tau)$ dependent on one variable has finite support, the function of two variables $\alpha(x+y)$ is not a function with finite support. This means that we must prove the existence of the functional $f * g$. It is known (see Gel'fand and Shilov, 1959), that for the functional $f * g$ to exist it is sufficient that one of the functionals should have finite support.

In tomography problems, we have the convolution of the function $1/x^2$ with the initial data, which are regular and have finite support. We can show also that the required convolution is expressed by the formula

$$\begin{aligned} S(r,\varphi) &= I(r,\varphi) * \left(-\frac{1}{\pi r^2}\right) \\ &= -\frac{1}{\pi}\int_0^\infty \frac{I(z-x,\varphi) + I(z+x,\varphi) + I(z,\varphi)}{x^2}\,\mathrm{d}x. \end{aligned} \tag{2.2.5}$$

In real situations the function $I(r,\varphi)$ is known on a certain discrete set of points. In order to use formula (2.2.5), we should construct an approximation of the function $I(r,\varphi)$ such that the integral in the right-hand side is meaningful. The integral (2.2.5) is convergent if the function $I(r, \varphi)$ belongs to the set K, i.e., if it has finite support and is infinitely differentiable.

However, the approximation of data by an infinitely differentiable function can be tedious when constructing numerical algorithms. Besides, the use of infinitely differentiable functions can lead to the smoothing of the boundaries of domains with densities which are sharply distinct. In order that the integral in (2.2.5) be convergent, it is sufficient that the function $I(r, \varphi)$ should have finite one-sided derivatives of the first order with respect to the variable r at every point. This allows us, in particular, to use cubic splines for constructing an approximation of the function $I(r, \varphi)$.

The basic operations on generalized functions used in tomography problems are convolution, differentiation, and the Fourier transform. The basic idea of definition of the operations is that certain properties of functionals specified by means of regular functions are taken as the basis for defining the corresponding operations on generalized functions which are linear functionals.

On this basis, we construct the notion of the convolution defined above. This method can be demonstrated most clearly when defining the operation of differentiation of generalized functions.

Suppose that the linear functional f is specified by a regular function $f(x)$ which has integrable derivative. The action of the derivative on the function $\alpha(x)$ from the space of the basic functions is as follows:

$$\int_{-\infty}^{\infty} f'(x)\alpha(x)\,\mathrm{d}x = -\int_{-\infty}^{\infty} f(x)\alpha'(x)\,\mathrm{d}x. \tag{2.2.6}$$

Here we have used term-by-term integration and the fact that $\alpha(x)$ is equal to zero outside a certain interval.

This property is taken as the basis when we define the derivative of a generalized function. Suppose we have a functional f. The functional defined by the equality $(f', \alpha) = -(f, \alpha')$ is called the derivative of the functional f. Since all the functions from the space of basic functions are infinitely differentiable, this definition is well-posed and generalized functions have derivatives of any order.

We now turn to the definition of the Fourier transform in the sense of generalized functions. In the above definitions, the elements of the space of basic functions are real-valued functions. When we define the Fourier transform, we consider complex-valued functions as the basic functions.

We suppose that K is the space of complex-valued basic functions (infinitely differentiable functions with finite support).

Any complex-valued locally-integrable function $f(x)$ corresponds to the functional

$$(f, \alpha) = \int \bar{f}(x)\alpha(x)\,\mathrm{d}x,$$

where $\bar{f}(x)$ is the complex conjugate of $f(x)$, $\alpha(x) \in K$.

The set of all linear continuous functionals in K forms the complex space of generalized functions K'. We denote by Z the set of functions which are Fourier transforms of the functions from K.

The Fourier transform of the element f from the space K is the functional g in the space Z acting as follows:

$$(g, \psi) = 2\pi(f, \alpha). \tag{2.2.7}$$

Here α is the element from K whose Fourier transform is ψ. This means that in order to derive the action of the functional g on the function $\psi(\lambda)$ from the space Z we need to do the following:

1) find the function $\alpha(x)$ from the space K whose Fourier transform is the function $\psi(\lambda)$;

2) find the action of the functional f on the function $\alpha(x)$.

The spaces of basic functions and functionals acting on them are chosen so that both steps can be realized.

It should be noted here that generalized functions and their Fourier transforms are defined as linear functionals in various basic spaces. In this case, functions from the set Z, where the Fourier transform acts, are not functions with finite supports, but they are infinitely differentiable. This fact allows us to conserve many useful properties of generalized functions.

We use the Fourier transform of homogeneous functions in inversion formulas of the ray transformation. These formulas are basic for algorithms of solving problems of three-dimensional tomography. The classical Fourier transform for such functions does not exist. In the formulas it is understood in the sense of generalized functions.

We now consider this problem from the viewpoint of constructing the corresponding numerical algorithms in three-dimensional space.

We recall the definition of the ray transformation given in the previous sections.

The function

$$(R_1^+ f)(\alpha, S) = \int_0^\infty f(t\alpha + S)\,\mathrm{d}t \tag{2.2.8}$$

is called the ray transformation of the function $f(x) = f(x_1, x_2, x_3)$. This is the integral of $f(x)$ along the ray coming from the point $S = (s_1, s_2, s_3)$ in the direction of the vector $\alpha = (\alpha_1, \alpha_2, \alpha_3)$.

As noted above, along with the function $(R_1^+ f)(\alpha, S)$ we consider the function

$$(R_1 f)(\alpha, S) = \int_{-\infty}^{\infty} f(t\alpha + S)\,\mathrm{d}t,$$

which is the integral along the entire line or, equivalently, the sum of integrals along both rays coming from the point S (in the directions α and $-\alpha$).

Both functions are homogeneous functions of degree -1. This means that we have the equalities

$$(R_1^+ f)(\alpha, S) = \frac{1}{|\alpha|}(R_1^+ f)\Big(\frac{\alpha}{|\alpha|}, S\Big),$$
$$(R_1 f)(\alpha, S) = \frac{1}{|\alpha|}(R_1 f)\Big(\frac{\alpha}{|\alpha|}, S\Big).$$

Note also that $(R_1 f)(\alpha, S)$ is an even function and $(R_1^+ f)(\alpha, S)$ is not even.

The notion of homogeneity of degree λ can be naturally extended to generalized functions if we take the equality $g(\gamma x) = \gamma^\lambda g(x)$ as the basic one. In terms of action on the basic function φ, this equality is written in the form $(g, \varphi(x/\gamma)) = \gamma^{\lambda+n}(g, \varphi(x))$. Here γ is any real number greater than zero, n is the dimension of the space where the basic functions are given. In the integral representation of generalized functions, the integer n appears after the corresponding change of variables in $\mathrm{d}x$.

It is known (Gel'fand and Shilov, 1959) that the Fourier transform of a homogeneous generalized function is also a homogeneous generalized function.

For integrable bounded functions f with bounded support, their ray transformation is a regular homogeneous function. From the results of Tuy (1983) and Trofimov (1991) it follows that in three-dimensional space the Fourier transform of such functions understood in the generalized sense is given by a regular function. A regular homogeneous function is specified by its values on the unit sphere. Thus, in practical situations, when we invert the ray transformation, we are interested in the connection between two functions. One of them is the restriction of the ray transformation to the unit sphere; the other one is the restriction of the Fourier transform of the ray data understood in the sense of generalized functions to the unit sphere. It is natural to call the similar transformation of the function given in the unit sphere the Semyanistyi transformation, since in the paper of Semyanistyi (1961) similar relations were first obtained for symmetric homogeneous functions in n-dimensional spaces. As was noted above, the function $(R_1^+ f)(\alpha, S)$ is not symmetric. The corresponding relations for the functions on the unit spheres in three-dimensional space were given in previous sections.

Previously we considered the inversion formulas of the ray transformation based on the explicit use of generalized functions and the methods which allow us to reduce these formulas to the form convenient for construction of numerical algorithms.

The formulas of inversion of the ray transformation can be obtained by the method which does not use generalized functions explicitly (see Smith, 1985). We shall show here that this method also essentially uses the Fourier transform of generalized functions.

In Smith (1985), the function

$$g(\beta, \Phi) = \int_{-\infty}^{\infty} f(\Phi + t\beta)\, \mathrm{d}t$$

is called the ray data. Here $\Phi = (\Phi_1, \Phi_2, \Phi_3) \in \mathbb{R}^3$, $\beta \in S^2$ (S^2 is the unit sphere). (It is easy to see that this function is $(R_1 f)(\beta, \Phi)$ in our notations).

The following functions are used in the inversion formula (see Smith, 1985):

$$H_\varepsilon = \begin{cases} 1/\varepsilon^2, & \text{if} \quad |t| < \varepsilon, \\ 1/t^2, & \text{if} \quad |t| \geq \varepsilon, \end{cases} \tag{2.2.9}$$

$$G(\xi, \Phi) = 2 \lim_{\varepsilon \to 0} \int_{S^2/2} H_\varepsilon(\xi \cdot \beta) g(\beta, \Phi)\, \mathrm{d}\beta, \tag{2.2.10}$$

where $S^2/2$ is a half of the unit sphere and (ξ, β) is the scalar product of the vectors β and ξ.

The inversion formulas in Smith (1985) are as follows:

$$f(x) = \frac{1}{4\pi^3} \int_{S^2/2} \lim_{\varepsilon \to 0} \int_{-R}^{R} F(\xi, l) H_\varepsilon((x, \xi) - l)\, \mathrm{d}l\, \mathrm{d}\xi, \tag{2.2.11}$$

where $2\pi F(\xi, (\Phi, \xi)) = G(\xi, \Phi)$, R is the radius of the ball containing the support of the function $f(x)$; $\mathrm{d}\xi$ is the surface element of the unit sphere.

If for any l such that $|l| < R$ and any $\beta \in S^2/2$ there exists a point Φ in the source trajectory such that $\Phi \cdot \beta = l$ (the Kirillov – Tuy conditions hold), then formula (2.2.11) can be used for determination of the function $f(x)$.

In Smith (1985) it is noted that the function F in the case of three-dimensional cone-beam tomography reconstruction plays a role analogous to that of the Fourier transform in two-dimensional tomography. This fact is not accidental.

Indeed, as is shown in Trofimov (1993a), the Fourier transform of the function $g(\beta, \phi)$ with respect to β in the sense of generalized functions has the form

$$G(\xi, \Phi) = \int_{S^2/2} \frac{g(\beta, \Phi)}{(\beta, \xi)^2} \, \mathrm{d}^2\beta. \tag{2.2.12}$$

The denominator in (2.2.12) can be equal to zero and (2.2.12) can be understood in the sense of generalized functions. In Trofimov (1995), the following statement is proved: If $f\varphi \in C^2$, then

$$\lim_{\varepsilon \to 0} \int_{-\infty}^{\infty} H_\varepsilon(t)\varphi(t) \, \mathrm{d}t = \int_0^{\infty} \frac{\varphi(t) + \varphi(-t) - 2\varphi(0)}{t^2} \, \mathrm{d}t = \Big(\frac{1}{t^2}, \varphi(t)\Big). \tag{2.2.13}$$

Taking (2.2.13), (2.2.12), and (2.2.10) into account, we see that the function $G(\xi, \Phi)$ is the Fourier transform of the function $g(\beta, \Phi)$ in the sense of generalized functions, and the function F in the inversion formula is defined by the function $G(\xi, \Phi)$.

2.3. THE RELATIONS BETWEEN THE RADON, FOURIER, AND RAY TRANSFORMATIONS

In the previous section we considered the formulas of inversion of the ray transformation. Also, there are methods of tomography reconstruction based on the preliminary calculation of the Fourier or Radon transform of the desired function. As was mentioned above, in the case of two variables, the ray transformation and the Radon transform coincide. In three-dimensional case these transformations differ.

In order to understand tomography methods, it is useful to obtain the relations between various forms of the transformations. Many of these relations can be obtained in the spaces of arbitrary dimensions. Here we shall consider only the cases important for practice, namely, two-dimensional and three-dimensional cases.

1. The relation between the Radon and Fourier transforms.

We suppose that $\tilde{f}(\lambda_1, \lambda_2, \lambda_3)$ is the Fourier transform of the function $f(x_1, x_2, x_3)$:

$$\tilde{f}(\lambda_1, \lambda_2, \lambda_3) = \int f(x_1, x_2, x_3) \exp\{\mathrm{i}(\lambda_1 x_1 + \lambda_2 x_2 + \lambda_3 x_3)\} \, \mathrm{d}x_1 \, \mathrm{d}x_2 \, \mathrm{d}x_3.$$

Integrating first along the plane $\lambda_1 x_1 + \lambda_2 x_2 + \lambda_3 x_3 = p$ for a fixed p, and then with respect to p, we obtain the well-known expression connecting the Fourier and Radon transforms

$$\tilde{f}(\lambda) = \int \check{f}(\lambda, p) \exp(\mathrm{i}\lambda p)\, \mathrm{d}p. \tag{2.3.1}$$

2. The relation between the Radon transform and the Fourier transform of the ray data.

In Gel'fand and Goncharov (1987), the method of inverting the ray transformation is proposed. In this method, from the initial data, we reconstruct the Radon transform of the function $f(x)$,

$$\check{f}(\xi, r) = \int f(x)\delta(r - (\xi, x))\, \mathrm{d}x,$$

which allows us to reconstruct $f(x)$ using well-known formulas.

The function

$$(S_x f)(\xi) = \int \frac{\check{f}(\xi, r)}{(r - (x, \xi))^2}\, \mathrm{d}r \tag{2.3.2}$$

is used in Gel'fand and Goncharov (1987) for deducing the inversion formulas.

We can show (see Trofimov, 1993a) that the following relation holds for the functions $(\tilde{R}_1^+ f)(\xi, x)$ and $(S_x f)(\xi)$:

$$\mathrm{Re}(\tilde{R}_1^+ f)(\xi, x) = C(S_x f)(\xi). \tag{2.3.3}$$

Here C is a certain constant. Equalities (2.3.2) and (2.3.3) represent the relation between the Radon transform and the ray transformation in three-dimensional space:

$$\mathrm{Re}(\tilde{R}_1^+ f)(\xi, x) = C \int \frac{\check{f}(\xi, r)}{(r - (x, \xi))^2}\, \mathrm{d}r. \tag{2.3.4}$$

Note that

$$(R_1 f)(\xi, x) = (R_1^+ f)(\xi, x) + (R_1^+ f)(-\xi, x)$$

yields

$$(\tilde{R}_1 f)(\xi, x) = 2\,\mathrm{Re}((\tilde{R}_1^+ f)(\xi, x)).$$

Equality (2.3.4) can be written in the form $2C(S_x f)(\xi) = (\tilde{R}_1 f)(\xi, x)$. From this equality and the definition of the function $(S_x f)(\xi)$ it follows that $(\tilde{R}_1 f)(\xi, x)$, as a function dependent on x, is constant on all the planes orthogonal to the vector ξ, since for all x from this plane the scalar product (x, ξ) is constant. This fact is the basis for the majority of the methods of inversion of the ray transformation. This statement is proved in Kirillov (1961) for the case of complex spaces. For real spaces, this statement is proved in Tuy (1983), Blagoveshchenskii (1986), and Finch (1985). This statement can be used for reconstruction of the function $(\tilde{R}_1 f)(\xi, x)$ at the points x of the domain D from the values of this function in the boundary.

3. The relation between the Fourier transform of the ray data and the Fourier transform of the desired function $f(x)$.

In Tuy (1983), the following equation is obtained:

$$(\tilde{R}_1^+ f)(\xi, x) = \int_0^\infty \rho \tilde{f}(\rho\xi) \exp(\mathrm{i}2\pi\rho(x, \xi)) \, \mathrm{d}\rho. \tag{2.3.5}$$

This is the connection between the Fourier transform of the ray data and the Fourier transform of the function f. The Fourier transform is understood in the sense of generalized functions. In order to use this formula for determining the function f, we must have the formulas for calculation of the generalized Fourier transform on the basis of the ray data. Such formulas were given in the previous sections.

Chapter 3.

Inverse kinematic problem in the tomographic setting

3.1. DIRECT KINEMATIC PROBLEM AND NUMERICAL SOLUTION FOR THREE-DIMENSIONAL REGULAR MEDIA

In seismology, the parameters characterizing the propagation of seismic waves are usually subdivided into dynamic and kinematic parameters. The dynamic parameters are amplitude, energy, the type of impulse, spectral characteristics, spatial polarization, and interference effects. The kinematic parameters are connected with studying wave fronts, rays, and with the measurements of the time of wave propagation. This classification is rather conventional, since when solving the dynamic problems and separating the real signal from background noise we have to use the kinematic wave characteristics. It should be noted that the kinematic characteristics are measured more precisely, for example, the time of impulse initiation in kinematics and the ground shifting amplitudes in dynamics (see Puzyrev (1992)).

By the direct kinematic problem we mean the problem of constructing seismic rays and calculating the times of the signal travel along these rays for a given speed distribution in the studied medium.

We shall be interested in this information from the viewpoint of solving inverse problems in Sections 3.4 and 3.5, where the initial information is represented by the times of arrival of refracted waves. Besides, the algorithms of solving the inverse problems in question are based on using kinematic data (the form of rays, the arrival times for the waves coming along these rays) for a medium with linear dependence of the signal speed on the depth.

The form of the rays in the medium is based on the Fermat principle (see Puzyrev (1992)), which states that the path of the signal coming from the source to the receiver is the extremal line for the Fermat functional. It has the form of the curvilinear integral

$$T(\vec{x}^0, \vec{x}^1) = \int_{\Gamma(\vec{x}^0, \vec{x}^1)} \frac{\mathrm{d}s}{V}. \tag{3.1.1}$$

Here $\mathrm{d}s$ is the element of the curve $\Gamma(\vec{x}^0, \vec{x}^1)$; $\vec{x}^0$ is the source coordinate; $\vec{x}^1$ is the receiver coordinate; $V = V(\vec{x})$ is the speed of the signal propagation in the medium; $\vec{x} \in R^3$; T is the time of the signal travel along $\Gamma(\vec{x}^0, \vec{x}^1)$.

We need to find a curve $\Gamma(\vec{x}^0, \vec{x}^1)$ (if $\vec{x}^0, \vec{x}^1$ and $V(\vec{x})$ are given) that provides the minimum of the functional (3.1.1). The formulated problem of variational calculus is known as the problem of geometric optics in space. Hereafter we shall use the solution of the direct kinematic problem with linear dependence of the signal propagation speed on the depth, namely

$$\begin{aligned} &V(z) = A + Bz, \qquad \vec{x} = (x, y, z) \\ &A = \mathrm{const}, \quad B = \mathrm{const} \\ &A > 0, \quad B > 0. \end{aligned} \tag{3.1.2}$$

Then the functional (3.1.1) can be written as follows:

$$T(\vec{x}^0, \vec{x}^1) = \int_{x_0}^{x_1} \frac{\sqrt{1 + y'^2 + z'^2}}{A + Bz} \, \mathrm{d}x,$$

where $\vec{x}_0 = (x_0, y_0, z_0)$, $\vec{x}_1 = (x_1, y_1, z_1)$, and the functions $y(x)$, $z(x)$ are determined from the Euler equation system. In our case, this system is

$$\begin{cases} y'' = 0 \\ z'' = -\dfrac{1 + y'^2 + z'^2}{A/B + z} \end{cases} \tag{3.1.3}$$

with the boundary conditions

$$\begin{aligned} y(x_0) = y_0, \quad z(x_0) = z_0, \\ y(x_1) = y_1, \quad z(x_1) = z_1. \end{aligned} \tag{3.1.4}$$

The first equation of system (3.1.3) defines the plane orthogonal to the plane $z = 0$.

The second equation, after transformations, becomes as follows:

$$R^2 = (z + A/B)^2 + (x + C_1)^2 + (y + C_2)^2.$$

This is a sphere.

Below we take $z_0 = z_1$. Taking into account the values of the constants determined from boundary conditions (3.1.4), we can write the system for the ray as follows:

$$\begin{cases} \left(z+\dfrac{A}{B}\right)^2 + \left(x - \dfrac{x_0 + x_1}{2}\right)^2 + \left(y - \dfrac{y_0 + y_1}{2}\right)^2 = R^2, \\ y \quad = \dfrac{y_1 - y_0}{x_1 - x_0}x + \dfrac{y_0 x_1 - y_1 x_0}{x_1 - x_0}. \end{cases} \tag{3.1.5}$$

The ray trajectory (the geodesic curve) is determined from system (3.1.5) as the trace of the intersection of the sphere (the first equation) and the plane orthogonal to the plane $z = 0$ (the second equation).

Suppose we observe the day surface, i.e., the plane $z = 0$, when $\vec{x}_0 = (x_0, y_0, 0)$, $\vec{x}_1 = (x_1, y_1, 0)$. We denote by d the distance between the source and the receiver. Then

$$d = \sqrt{(x_1 - x_0)^2 + (y_1 - y_0)^2}.$$

In order to calculate the time $T(\vec{x}^0, \vec{x}^1)$, we can use the formula

$$T(\vec{x}^0, \vec{x}^1) = \frac{1}{B} \ln \frac{2R + d}{2R - d}. \tag{3.1.6}$$

In the cases when analytic solution of the direct problem is impossible, we need to solve it by numerical methods.

It should be noted that if the medium in question admits partition into the domains where the speed of signal propagation can be approximated by a linear function or by a constant, then we can use the parts of circles or lines and calculate the signal travel time by the corresponding formulas. However, such partition is not always possible. For example, in a medium where the speed changes continuously, there is a serious problem connected with imaginary boundaries within the medium.

If the speed in the medium changes so that we cannot obtain an analytic expression for the ray and the time of travel along this ray, then the direct kinematic problem must be solved numerically (see Psencik, 1983; Cerveny, Molotkov, and Psencik, 1977).

In order to solve the direct kinematic problem numerically we need to specify the distribution of the speed of the signal propagation in the medium and specify the way of constructing the ray trajectories.

We shall assume that the representation of the speed in the medium is given by a function of three variables $V(x, y, z)$ sufficiently smooth for the needs of further argument and show how to construct these rays. The basic equation of kinematic seismology is the eikonal equation

$$(\nabla T(x, y, z))^2 = \frac{1}{V^2(x, y, z)}.$$

The rays have the following well-known property: they are the characteristics of the eikonal equation. We suppose that a certain ray trajectory is chosen and we can calculate the distance along this trajectory

$$\vec{x} = \vec{x}(s).$$

Here $\vec{x}$ is the radius vector of a point of the ray; s is the distance along this ray. The vector $\vec{x}$ is represented as follows:

$$\vec{x} = x(s)\vec{i} + y(s)\vec{j} + z(s)\vec{k},$$

and the basis vector $\vec{t}$ tangent to the ray is given by

$$\vec{t} = \frac{\mathrm{d}\vec{x}}{\mathrm{d}s} = \dot{x}(s)\vec{i} + \dot{y}(s)\vec{j} + \dot{z}(s)\vec{k}.$$

Since the eikonal equation for every s is an identity holding along the ray, we can differentiate it with respect to s:

$$\frac{\mathrm{d}}{\mathrm{d}s}\left(\nabla T\right) = \frac{\mathrm{d}}{\mathrm{d}s}\left(\frac{\vec{t}}{V}\right),$$

or

$$\nabla\frac{\mathrm{d}T}{\mathrm{d}s} = \frac{\mathrm{d}}{\mathrm{d}s}\frac{\vec{t}}{V},$$

$$\frac{\mathrm{d}}{\mathrm{d}s}\left(\frac{\vec{t}}{V}\right) = \nabla\frac{1}{V}. \tag{3.1.7}$$

This equation yields three differential equations of the second order.

The derivative and the gradient in (3.1.7) are represented as follows:

$$\frac{1}{V}\frac{\mathrm{d}\vec{t}}{\mathrm{d}s}-\vec{t}\frac{1}{V^2}\frac{\mathrm{d}V}{\mathrm{d}s}=-\frac{1}{V^2}\left(\frac{\partial V}{\partial x}\vec{i}+\frac{\partial V}{\partial y}\vec{j}+\frac{\partial V}{\partial z}\vec{k}\right). \tag{3.1.8}$$

Since we have

$$\frac{1}{V}\frac{\partial V}{\partial x}=\frac{\partial \ln V}{\partial x},$$

$$\frac{\partial \ln V}{\partial s}=\frac{\partial \ln V}{\partial x}\dot{x}+\frac{\partial \ln V}{\partial y}\dot{y}+\frac{\partial \ln V}{\partial z}\dot{z},$$

$$\frac{\mathrm{d}\vec{t}}{\mathrm{d}s}=\ddot{x}(s)\vec{i}+\ddot{y}(s)\vec{j}+\ddot{z}(s)\vec{k},$$

we obtain

$$\frac{1}{V}\ddot{x}-\frac{1}{V}\dot{x}=-\frac{1}{V}\frac{\partial \ln V}{\partial x}.$$

For the variables y and z, we have analogous expressions. Taking into account that

$$\frac{1}{V^2}\frac{\partial V}{\partial s}=\frac{1}{V}\frac{\partial \ln V}{\partial s}=\frac{1}{V}\left(\frac{\partial \ln V}{\partial x}\dot{x}+\frac{\partial \ln V}{\partial y}\dot{y}+\frac{\partial \ln V}{\partial z}\dot{z}\right),$$

we can find the projection of the vector equality (3.1.8) onto the axes x, y, z:

$$\frac{1}{V}\ddot{x}-\frac{1}{V}\dot{x}\left(\frac{\partial \ln V}{\partial x}\dot{x}+\frac{\partial \ln V}{\partial y}\dot{y}+\frac{\partial \ln V}{\partial z}\dot{z}\right)+\frac{1}{V}\frac{\partial \ln V}{\partial x}=0,$$

$$\frac{1}{V}\ddot{y}-\frac{1}{V}\dot{y}\left(\frac{\partial \ln V}{\partial x}\dot{x}+\frac{\partial \ln V}{\partial y}\dot{y}+\frac{\partial \ln V}{\partial z}\dot{z}\right)+\frac{1}{V}\frac{\partial \ln V}{\partial y}=0,$$

$$\frac{1}{V}\ddot{z}-\frac{1}{V}\dot{z}\left(\frac{\partial \ln V}{\partial x}\dot{x}+\frac{\partial \ln V}{\partial y}\dot{y}+\frac{\partial \ln V}{\partial z}\dot{z}\right)+\frac{1}{V}\frac{\partial \ln V}{\partial z}=0.$$

Dividing by $1/V$, we finally obtain

$$\begin{cases} \ddot{x}-(\dot{x}^2-1)\dfrac{\partial \ln V}{\partial x}-\dfrac{\partial \ln V}{\partial y}\dot{x}\dot{y}-\dfrac{\partial \ln V}{\partial z}\dot{x}\dot{z}=0, \\[2ex] \ddot{y}-(\dot{y}^2-1)\dfrac{\partial \ln V}{\partial y}-\dfrac{\partial \ln V}{\partial x}\dot{x}\dot{y}-\dfrac{\partial \ln V}{\partial z}\dot{y}\dot{z}=0, \\[2ex] \ddot{z}-(\dot{z}^2-1)\dfrac{\partial \ln V}{\partial z}-\dfrac{\partial \ln V}{\partial x}\dot{x}\dot{z}-\dfrac{\partial \ln V}{\partial y}\dot{z}\dot{y}=0. \end{cases} \tag{3.1.9}$$

In such a form, system (3.1.9) for the case of two variables was obtained by Belonosova and Alekseev (1967). System (3.1.9) can be resolved for the highest derivatives. This makes this system convenient for numerical solution:

$$
\begin{aligned}
\ddot{x} &= \frac{\partial \ln V}{\partial y}\dot{x}\dot{y} + \frac{\partial \ln V}{\partial z}\dot{x}\dot{z} + (\dot{x}^2 - 1)\frac{\partial \ln V}{\partial x}, \\
\ddot{y} &= \frac{\partial \ln V}{\partial x}\dot{x}\dot{y} + \frac{\partial \ln V}{\partial z}\dot{y}\dot{z} + (\dot{y}^2 - 1)\frac{\partial \ln V}{\partial y}, \\
\ddot{z} &= \frac{\partial \ln V}{\partial x}\dot{x}\dot{z} + \frac{\partial \ln V}{\partial y}\dot{y}\dot{z} + (\dot{z}^2 - 1)\frac{\partial \ln V}{\partial z}.
\end{aligned}
\tag{3.1.10}
$$

If we specify the initial data

$$
\begin{aligned}
x|_{s=s_0} &= x^0, \quad y|_{s=s_0} = y^0, \quad z|_{s=s_0} = z^0, \\
\dot{x}|_{s=s_0} &= a, \quad \dot{y}|_{s=s_0} = b, \quad \dot{z}|_{s=s_0} = c,
\end{aligned}
\tag{3.1.11}
$$

then, solving the Cauchy problem (3.1.10)–(3.1.11), we construct the ray coming from the point (x^0, y^0, z^0) with the direction cosines a, b, and c. The speed $V(x, y, z)$ is assumed to be known.

To calculate the times of travel along the desired ray, we add one more equation to system (3.1.10):

$$\dot{T} = V^{-1},$$

and the condition

$$T|_{s=s_0} = T_0.$$

is added to the initial data (3.1.11).

Because system (3.1.10) is resolved for the highest derivatives, we can introduce the additional variables $\tilde{x} = \dot{x}$, $\tilde{y} = \dot{y}$, and $\tilde{z} = \dot{z}$ and reduce the system of three differential equations to the system of six differential equations of the first order with initial data (3.1.11) transformed in the corresponding way. This makes the problem convenient for application of the difference methods. The most well-known method for numerical solution of the direct problem is the Runge–Kutta method. We decided to choose this method for our research.

The main role in the accuracy of this method in constructing the ray trajectory is played by the mesh spacing H in calculation. It is natural that, with smaller spacing, the ray constructed using this method is closer to the real ray. However, as the spacing H becomes smaller, the volume of computation increases. Besides, in the general case, if constant spacing is used in construction of the ray, then the specific character of the ray behavior is not taken into account. Choosing a very small spacing leads to the difficulties mentioned above. So, the question arises of how to choose an optimal spacing in the calculation process.

To this end, in practice, we usually apply the calculating scheme with automatic choice of the spacing H. Specifically, we do double calculation for every point, first, with the spacing H, then with the spacing $H/2$. If the obtained results differ to within the given accuracy ε, then for further computation we take the spacing equal to $H/4$, and the process is repeated. The process is finished when the accuracy is within the given accuracy.

Thus, given the speed distribution in the medium, we may construct the ray trajectories starting from the point where the point source is situated and calculate the travel time along the constructed ray.

The direct kinematic problem in the two-point setting consists in constructing the ray which connects the source and the receiver and calculating the time of the wave travel along this ray. This problem is more complicated than the problem with initial data (the Cauchy problem) for the system of ray equations considered above. Indeed, when we solve the Cauchy problem, we may construct a ray and calculate the time of signal travel along this ray. But it is not known whether the source and the fixed receiver will be connected by the given ray.

For solving the two-point problem, we have used the following iterative approach: the rays and the arrival times are defined by the solution of the problem with initial data. However, the direction cosines of the ray at the point of generation are chosen successively, so that each ray passes closer to the receiver than the previous one. Such an approach to solution of the boundary-value problem is called the "adjustment" method. The problem of constructing the desired sequence $\{a_i, b_i, c_i\}$ arises.

The most well-known and effective method for solving two-dimensional problems is the iterative process of choosing initial data based on dividing the interval in two (see Psencik (1983)). In three-dimensional case, the problem can be simplified a little, since the angles at which the ray arrives vary from 0 to $\pi/2$ and the relation $a^2 + b^2 + c^2 = 1$ allows us to express one cosine in terms of the other two cosines. However, this method is rather tedious for computer realization.

Making use of the specific character of the problem, we can construct an algorithm that makes it possible to avoid these difficulties.

We now consider the observation system (the disposition of the sources and receivers) applied to obtain the information used for solving the inverse kinematic problem. This observation system is the circle

$$x^2 + y^2 = r^2, \tag{3.1.12}$$

where the sources and receivers are concentrated.

When we solve the inverse problem, we consider only regular media with quasilinear dependence of the speed on the variable z:

$$V(\vec{x}) = V_0(z) + V_1(\vec{x}), \quad \vec{x} = (x, y, z), \tag{3.1.13}$$

$$V_0 \gg V_1, \quad V_0(z) = A + Bz, \tag{3.1.14}$$

$$A = \text{const}, \quad B = \text{const}, \quad A > 0, \quad B > 0.$$

Being slightly different from the signal travel time $T_0(\vec{x}_0, \vec{x}_1)$ in the medium with the speed of signal propagation $V_0(z)$, the function $T(\vec{x}_0, \vec{x}_1)$ conserves the necessary smoothness. Moreover, we shall assume further that the sources and the receivers are sufficiently closely spaced. This provides the smooth change of the component T_i, $i = \overline{1, N}$ in the vector of the measured data; N is the number of the source-receiver pairs.

Taking all this into account, we organized the following calculating scheme of solution of the two-point direct kinematic problem. We denote by $R(j, g)$ the distance from the point at which the ray comes to the surface of observation to the circle where our observation system is located. We denote by j the angle between the X-axis and the line which connects the source and the receiver. We know this angle since we know the coordinates of the source and the receiver. The angle g is the angle between the tangent direction to the ray at the point of the signal generation and the coordinate axis z:

$$g = \arccos \dot{z}|_{s=0}.$$

We consider some specific source-receiver pair. In this case, the angle j is fixed. Denote it by j^*. Set

$$R(j^*, g) = R(g).$$

We consider the following auxiliary problem: Find an angle g^* such that

$$R(g^*) < \varepsilon, \tag{3.1.15}$$

where ε is the given accuracy.

The left-hand side of inequality (3.1.15) is given implicitly. However, for various g we must find a ray such that the point at which it intersects the plane $z = 0$ is situated in an ε-neighborhood of our circle. To this end, we shall organize the iterative process of determining g^*.

We consider the equation

$$R(g) = 0. \tag{3.1.16}$$

Since we consider only regular media and the source is fixed as well as the angle j^*, equation (3.1.16) has a unique root.

We shall solve this equation numerically by the Newton method. The recurrent formula for the root of equation (3.1.16) in the Newton method is as follows:

$$g_i = g_{i-1} - \frac{R(g_{i-1})}{R'(g_{i-1})}, \tag{3.1.17}$$

where i is the number of iteration.

We replace the derivative $R'(g_{i-1})$ by the difference relation

$$R'(g_{i-1}) = \frac{R(g_{i-1} + \Delta g_i) - R(g_{i-1})}{\Delta g_i}, \tag{3.1.18}$$

where $\Delta g_i = g_i - g_{i-1}$. Taking into account (3.1.18), we rewrite the recurrent formula (3.1.17) in the form

$$g_i = \frac{(g_{i-1} + \Delta g_i)R(g_{i-1}) - g_{i-1}R(g_{i-1} + \Delta g_i)}{R(g_{i-1}) - R(g_{i-1} + \Delta g_i)}.$$

In order to refine the convergence of the process, we have used variable spacing Δg_i. After each iteration, Δg_i was divided by two. For the initial approximation $g = g_0$ we take the angle corresponding to the given two-point problem in the case of the medium with the speed given by (3.1.15). The angle g_0 is obtained easily from the geometric considerations:

$$g_0 = \arctan \frac{2A}{Bd}.$$

Systematic errors in calculation of $R(g)$ caused by approximate calculations determine the restrictions imposed on the value of Δg. We set

$$\Delta R(\Delta g) = |R(g + \Delta g) - R(g)|.$$

The values of $\Delta R(\Delta g)$ must exceed the value of possible noise by an order of magnitude.

We note one more specific feature of calculations which arises when we solve the two-point problem. There are errors caused by the way of fixing the point at which the ray comes to the plane $z = 0$. The reason is that the criterion for the ray to arrive at the surface $z = 0$ is $z_k \leq 0$, where k is the step number in the Runge–Kutta iterative process. This natural criterion introduces an error in the calculation of $R(g)$ and of the time of travel along this ray. To reduce this error when approaching the surface of observation, i.e., when z_i becomes less than a certain given value z', we choose the spacing H sufficiently small and constant in order to fix the time of arrival of the ray at the surface more precisely.

Solving the auxiliary problem, we have found the way by means of which we can "drop" three points in the neighborhood of the receiver and calculate the times of the signal arrival at these points. Indeed, taking the angles $j_1 = j^* - \Delta j$, $j_2 = j^*$, and $j_3 = j^* + \Delta j$ for the angle j and solving the auxiliary problem for them, we obtain the times of arrival of the signal at three points in the neighborhood of the receiver (Δj is sufficiently small). We denote these times by T_1, T_2, and T_3, respectively. Further, using the smoothness of the function $T(\vec{x}_0, \vec{x})$, we can do the interpolation of the values T_1, T_2, and T_3 at the point where the receiver is situated.

We now show the results of computations which illustrate the solution of the three-dimensional two-point direct kinematic problem following the developed algorithm.

We use the explicit formula for $T(\vec{x}_0, \vec{x})$ presented in Lagunova and Omel'chenko (1981) for the media with the speed

$$V(\vec{x}) = \left(\sqrt{\lambda_0^2 + \langle \beta, \vec{x} \rangle + \alpha^2 |\vec{x}|^2}\right)^{-1}. \tag{3.1.19}$$

Here $\vec{x} \in \mathbb{R}^3$, β is a constant three-dimensional vector, $\langle \cdot, \cdot \rangle$ is the scalar product, λ_0 is a constant.

For such media, the following explicit formula for $T(\vec{x}_0, \vec{x})$ is obtained by Lagunova and Omel'chenko (1981):

$$\begin{aligned} T(\vec{x}_0, \vec{x}) = {} & \frac{M\sqrt{M^2 - 1}}{2\alpha^3}(\alpha^2(\lambda_0^2 + 2\langle \beta, \vec{x}_0 \rangle + \alpha^2 |\vec{x}_0|^2) + (\beta + \alpha^2 \vec{x}_0)^2) \\ & + \frac{\sqrt{M^2 - 1}}{\alpha^3}((\alpha^2 \vec{x}_1 + \beta) - (\alpha^2 \vec{x}_0 + \beta)M)(\beta + \alpha^2 \vec{x}_0) \qquad (3.1.20) \\ & + \frac{\ln(M + \sqrt{M^2 - 1})}{2\alpha^3}(\alpha^2(\lambda_0^2 + 2\langle \beta, \vec{x}_0 \rangle + \alpha^2 |\vec{x}_0|^2) - (\beta + \alpha^2 \vec{x}_0)^2), \end{aligned}$$

where

$$M = \frac{-(\alpha^2\vec{x}_1 + \beta)(\beta + \alpha^2\vec{x}_0)}{\alpha^2\lambda_0^2 - \beta^2} + \frac{\sqrt{Q}}{\alpha^2\lambda_0^2 - \beta^2}$$

$$Q = [(\alpha^2\vec{x}_1 + \beta)(\beta + +\alpha^2\vec{x}_0)]^2 + (\alpha^2\lambda_0^2 - \beta^2)[\alpha^2(\lambda_0^2 + 2\langle\beta, \vec{x}_1\rangle + \alpha^2|\vec{x}_1|^2) + (\alpha^2\vec{x}_0 + \beta)].$$

We set

$$V(x, y, z) = \sqrt{1 - z + 2 \cdot 10^{-4}(y + x) + (0.51)^2(x^2 + y^2 + z^2)}.$$

The source will be situated at the point $(-1, 0, 0)$ and the receiver at the point $(1, 0, 0)$. Table 3.1 represents the results of dropping three points in the receiver neighborhood. In this case $j^* = 0$, and the accuracy is equal to $\varepsilon = 10^{-4}$. The difference between the arrival time calculated according to the scheme proposed above and by the explicit formula (3.1.20) is as follows:

$$2.0039690 - 2.003095 = 0.874 \cdot 10^{-3}.$$

The relative error is equal to $0.874 \cdot 10^{-3}/2.003969 \approx 0.05\%$.

When solving the inverse problem, we used the difference $T - T_0 = \Delta T$. Here T_0 is the time for the medium specified by (3.1.15). If $V_0(z) = 1 + 0.5\,z$, then $T_0 \approx 1.92$ and $\Delta T \approx 1\%$. Note that the longest ray trajectory corresponds to the source-receiver pair for our observation system. All further computations will be done for the case $r = 1$ in formula (3.1.12), i.e., the observation system will be considered only in the circle of the unit radius.

The calculations illustrate the accuracy of the solution of the direct two-point kinematic problem by the developed numerical algorithm. They are important for evaluation of the accuracy of solution of the inverse kinematic problem when we study the noise stability of the algorithm and its sensitivity to systematic and random errors in the initial data.

The number of the point	The number of the iteration	φ	R	T	The result of interpolation	The time calculated by the explicit formula
1	6	-1.8	0.00000133	2.002994		
2	3	0	-0.00007034	2.001294	2.003095	2.003969
3	3	1.8	-0.00007302	1.996394		

Table 3.1

3.2. FORMULATION OF THE INVERSE KINEMATIC PROBLEM WITH THE USE OF A TOMOGRAPHIC SYSTEM OF DATA GATHERING

We consider a three-dimensional inhomogeneous medium with index of refraction $n(x,y,z) = V^{-1}(x,y,z)$, where $V(x,y,z)$ is the speed of propagation of vibrations in the medium. We generate the signal at the point $S_0 = (x_0, y_0, 0)$ and register the time of arrival of the refracted wave $T(S_0, S_1)$ at the point $S_1 = (x_1, y_1, 0)$. The inverse kinematic problem consists in determining the function $V(x,y,z)$, provided that the function $T(S_0,S_1)$ is given. In the general case, the inverse kinematic problem is overdetermined: given the function of four variables $T(x_0,y_0,x_1,y_1)$, it is required to determine the function $V(x,y,z)$ of three variables. The inverse kinematic problem in the overdetermined formulation was investigated by Romanov and Mukhometov.

The observation system in the form of the circle of radius r allows us to avoid overdetermination of the problem, because in this case T is the function dependent on the radius r and two angles φ_1 and φ_2 (in polar coordinates) on the source and the receiver, respectively. This observation system was used when solving the direct problem in Section 3.1, and it will be used further.

The medium in question is assumed to be regular. This means that the change of speed in the medium is such that any pair of source-receiver points there corresponds to one geodesic curve (ray) $\Gamma(S_0,S_1)$. The other important assumption is that the speed $V(x,y,z)$ is represented in the form (3.1.14)–(3.1.15).

The relation between the values A and B provides sufficiently deep travel of the ray for the given base of observation (the distance between the source and the receiver).

The function $V_0(z)$ is known, i.e., the numbers A and B are given.

In what follows, we shall use the linearization method for solving the inverse kinematic problem for multidimensional media, which is often used following Romanov (1987) (see also the bibliography therein).

We now write the eikonal equation

$$[\operatorname{grad} T(S_0,S_1)]^2 = n^2(x,y,z). \tag{3.2.1}$$

The functions $T(S_0,S_1)$ and $n(x,y,z)$, by virtue of (3.1.14), become as follows:

$$\begin{aligned} T(S_0,S_1) &= T_0(S_0,S_1) + \varepsilon T_1(S_0,S_1), \qquad 0 < \varepsilon \ll 1 \\ n(x,y,z) &= n_0(z) + \varepsilon n_1(x,y,z), \qquad n_0(z) = (A+Bz)^{-1}. \end{aligned} \tag{3.2.2}$$

Substituting expressions (3.2.2) into (3.2.1), we obtain

$$[\mathrm{grad}(T_0 + \varepsilon T_1)]^2 = (n_0 + \varepsilon n_1)^2(x, y, z),$$

or

$$|\nabla T_0|^2 + 2\varepsilon\langle \nabla T_0, \nabla T_1\rangle + \varepsilon^2 |\nabla T_1|^2 = n_0^2 + 2\varepsilon n_0 n_1 + \varepsilon^2 n_1^2.$$

Omitting the terms of order ε^2, we obtain

$$|\nabla T_0|^2 + 2\varepsilon\langle \nabla T_0, \nabla T_1\rangle = n_0^2 + 2\varepsilon n_0 n_1.$$

Equating the coefficients of the same powers of ε, we have

$$|\nabla T_0|^2 = n_0^2,$$

or

$$|\nabla T_0| = n_0.$$

Finally, we have

$$\langle \nabla T_0, \nabla T_1\rangle = n_0 n_1.$$

Because the vector ∇T_0 is directed along the tangent direction to the ray $\Gamma(S_0, S_1)$ (the ray which corresponds to the medium with the speed $V = V_0$), we can write

$$\langle |\nabla T_0| \vec{e}, \nabla T_1\rangle = n_0 n_1,$$

where $\vec{e}$ is the unit vector tangent to Γ_0; furthermore

$$|\nabla T_0| \langle \vec{e}, \nabla T_1\rangle = n_0 n_1. \tag{3.2.3}$$

Dividing the left-hand side of equality (3.2.3) by $|\nabla T_0|$ and the right-hand side by n_0, we obtain

$$\langle \vec{e}, \nabla T_1\rangle = n_1. \tag{3.2.4}$$

The expression in the left-hand side of equality (3.2.4) is the derivative in the direction tangent to Γ_0. Integrate both sides of equality (3.2.4) along the ray Γ_0:

$$\int_{\Gamma_0(S_0,S_1)} \langle \vec{e}, \nabla T_1\rangle \,\mathrm{d}s = \int_{\Gamma_0(S_0,S_1)} n_1 \,\mathrm{d}s.$$

As a result, we obtain the formula

$$T_1(S_0, S_1) = \int_{\Gamma_0(S_0,S_1)} n_1 \,\mathrm{d}s. \tag{3.2.5}$$

Recall that $T_1 = T - T_0$. The values of T are known (the result of solution of the direct problem; in practice, it is the vector of the measured data), and the values of T_0 in the case $V_0 = A + Bz$ are derived explicitly. For this reason, the calculation of T_1 is not tedious.

At this stage of solution the problem is reduced to obtaining the function n_1 from the integral equation (3.2.5), provided that the function T_1 is known.

The use of the observation system in the form of a circle not only makes the problem determined. It also forms the tomographic setting of the problem in question. The rays Γ_0 stretched on the circle of observation system form the surface of ball segments (see Figure 3.1). By varying the radius r we obtain the system of imbedded ball segments filling the volume of the domain in $\mathbb{R}^3$. Determining n_1 on the surface of such ball segments under study we obtain the solution of three-dimensional problem. Note that the section-by-section study of the object and the method of acquisition of the projection data makes our problem similar to the known problem of classical tomography.

3.3. DEDUCTION OF THE BASIC INVERSION FORMULA AND THE ALGORITHM OF SOLVING THE INVERSE KINEMATIC PROBLEM IN THREE-DIMENSIONAL LINEARIZED FORMULATION

We suppose that the function $T_1(S_0, S_1)$ is known and (3.2.5) holds, where $n_1(x,z) = \dfrac{1}{V(x,z)} - \dfrac{1}{V_0(z)}$, $x \in \mathbb{R}^2$ and $V_0(z)$ is defined as in (3.1.15).

The points S_0 and S_1 are situated on the circle (3.1.12) ($x_1 = x$, $x_2 = y$). The ray trajectory $\Gamma(S_0, S_1)$ is derived from the system

$$\begin{cases} |x|^2 + \left(z + \dfrac{A}{B}\right)^2 = \rho^2, \\ p = \text{const}, \ \rho = \text{const}, \ \rho > 0, \\ \langle x, \nu \rangle - p = 0 = (\sin\theta, -\cos\theta). \end{cases} \tag{3.3.1}$$

The geometric meaning of the values p, ρ and θ is shown in Figure 3.2. In expression (3.2.5) we pass from integration along $\Gamma(S_0, S_1)$ to integration along $l(S_0, S_1)$. We denote

$$z = -\frac{A}{B} \pm \sqrt{\rho^2 - |x|^2} = z^*(x).$$

We shall consider the case $z \geq 0$, where the plus sign before the root should be chosen. For $\mathrm{d}s$ we have (see Figure (3.2)):

$$\mathrm{d}s = \mathrm{d}l\sqrt{1 + (z_l)^2}.$$

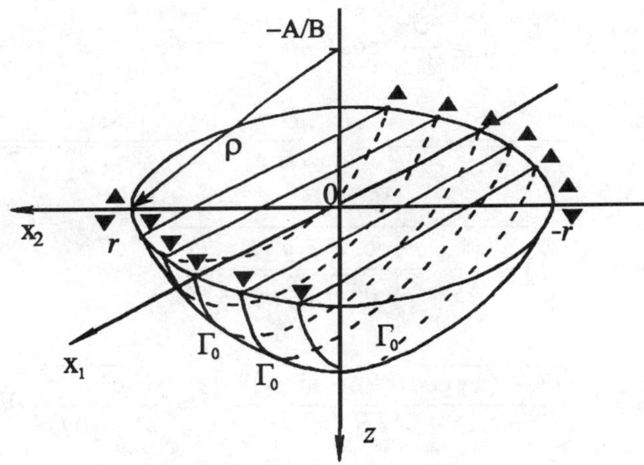

Figure 3.1: The system of observation and the surface formed by the rays in the case of linear dependence of the speed of the refracted wave on the depth (the variable z), 1 — sources, 2 — receivers of the signal.

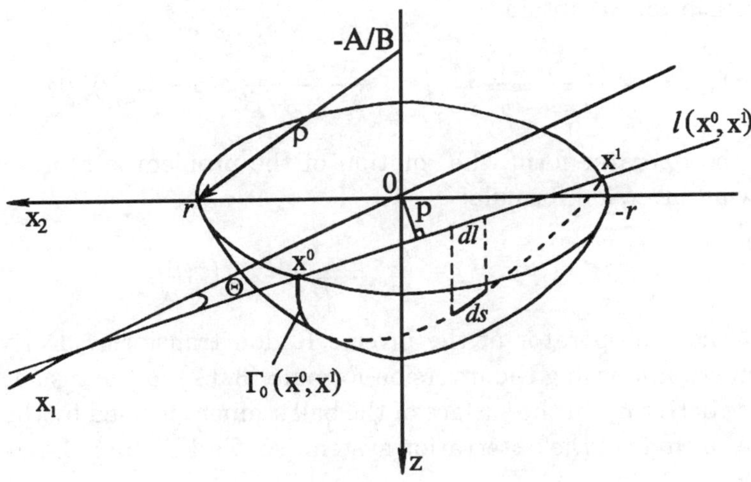

Figure 3.2

Since the equation of the line $l(S_0, S_1)$ is $x_1 \cos\theta - x_2 \sin\theta - p = 0$, we have

$$z_l = \frac{\partial z}{\partial x_1} \cos\theta + \frac{\partial z}{\partial x_2} \sin\theta,$$

$$\frac{\partial z}{\partial x_1} = \frac{x_1}{\sqrt{\rho^2 - x_1^2 - x_2^2}}, \qquad \frac{\partial z}{\partial x_2} = -\frac{x_2}{\sqrt{\rho^2 - x_1^2 - x_2^2}},$$

$$\mathrm{d}s = \mathrm{d}l \sqrt{1 + \Big(- \frac{x_1 \cos\theta + x_2 \sin\theta}{\sqrt{\rho^2 - x_1^2 - x_2^2}} \Big)^2}$$

$$= \frac{\sqrt{\rho^2 - (x_1 \cos\theta - x_2 \sin\theta)^2}}{z^* + A/B}\, \mathrm{d}l = \frac{\sqrt{\rho^2 - p^2}}{z^* + A/B}\, \mathrm{d}l.$$

Substituting the expressions for $\mathrm{d}s$ into (3.2.5), we obtain

$$T_1(S_0, S_1) = \int_{l(S_0,S_1)} n_1(x, z^*) \frac{\sqrt{\rho^2 - p^2}}{z^* + A/B}\, \mathrm{d}l,$$

or

$$\frac{T_1(S_0, S_1)}{\sqrt{\rho^2 - p^2}} = \int_{l(S_0,S_1)} \frac{n_1(x, z^*)}{z^* + A/B}\, \mathrm{d}l.$$

Passing from the last relation to the canonical representation of the integral Radon transform, we obtain

$$f(p, \theta) = \frac{T_1(S_0, S_1)}{\sqrt{\rho^2 - p^2}} = \int_{|x| \le r} \frac{n_1(x, z^*)}{z^* + A/B}\, \delta(p - \langle x, \nu \rangle)\, \mathrm{d}x.$$

Thus, in the operator form, the solution of the problem is represented by the following inversion formula:

$$n_1(x_1, x_2, z^*) = \Big(z^* + \frac{A}{B}\Big) R^{-1} f(p, \theta), \tag{3.3.2}$$

where R^{-1} is the operator of the inverse Radon transform. Deriving the desired function n_1 using the inversion formula (3.3.2) in the disk $x^2 + y^2 \le r^2$, we thus derive n_1 on the surface of the ball segment formed by the rays Γ_0 based on the circle of the observation system (see Zerkal, 1988; Lavrent'ev *et al.*, 1995).

We now show that the deduction of the similar formula is possible if and only if V_0 is a linear function, which corresponds to the fact that the rays Γ_0 are of the arc form.

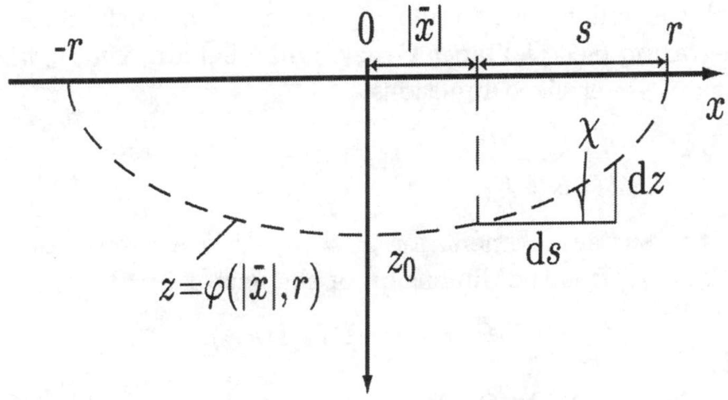

Figure 3.3

Suppose now that the ray trajectory Γ_0 is a smooth curve defined by a certain function φ. Then φ is determined from the equation

$$|x| = r - n_0(z_0)\int_0^{\varphi} \frac{du}{\sqrt{n_0^2(u) - n_0^2(z_0)}}. \tag{3.3.3}$$

Indeed, let us introduce the variable $s = r - |x|$, $ds = dz/\tan\chi$. Because $n_0(z)\cos\chi = n_0(0)\cos\chi_0$ (by the Snellius law), we obtain $\cos\chi = n_0(0)\cos\chi_0/n_0(z)$; furthermore, since $\chi = 0$ at the point z_0, we obtain $n_0(z_0) = n_0(0)\cos\chi_0$. Therefore, we have

$$\tan\chi = \frac{\sqrt{1 - n_0^2(0)\cos^2\chi_0/n_0^2(z)}}{n_0\cos\chi_0/n_0(z)} = \frac{\sqrt{n_0^2(z) - n_0^2(z_0)}}{n_0(z_0)}.$$

Hence

$$\int_0^z ds = n_0(z_0)\int_0^z \frac{du}{\sqrt{n_0^2(u) - n^2(z_0)}} = n_0(z_0)\int_0^{\varphi(|x|,r)} \frac{du}{\sqrt{n_0^2(u) - n^2(z_0)}},$$

$$s(\varphi) = n_0(z_0)\int_0^{\varphi(|x|,r)} \frac{du}{\sqrt{n_0^2(u) - n^2(z_0)}},$$

which is the desired conclusion.

In particular, for $n_0(z) = (A + Bz)^{-1}$, we have

$$\varphi = (\rho^2 - |x|^2)^{1/2} - \frac{A}{B}, \qquad \rho^2 = r^2 + \frac{A^2}{B^2}. \tag{3.3.4}$$

We now use the following formula of passing from surface integration to volume integration (see Gel'fand, Graev, and Vilenkin, 1966), which is used in the theory of generalized functions:

$$\int_{\vec{P}=0} \frac{u\,\mathrm{d}\sigma}{|\mathrm{D}\vec{P}|} = \int \delta(\vec{P})u\,\mathrm{d}x, \qquad x \in \mathbb{R}^n.$$

Here $\mathrm{d}\sigma$ is the surface element for $\vec{P} = 0$; $\vec{P}$ is a vector-function: $\vec{P} = (P_1(x), \dots, P_k(x))$; k is the dimension of the vector function,

$$|\mathrm{D}\vec{P}| = \det(\langle \mathrm{D}P_i, \mathrm{D}P_j\rangle),$$

$$\mathrm{D}P_i = (\mathrm{D}_{x_1}P_i, \mathrm{D}_{x_2}P_i, \dots, \mathrm{D}_{x_n}P_i), \qquad \mathrm{D}_{x_j} = \frac{\partial}{\partial x_j}.$$

In our case,

$$\begin{aligned} &\vec{P} = (P_1, P_2), \\ &P_1 = \langle x, \nu\rangle - p = 0, \\ &P_2 = z - \varphi(|x|, r) = 0, \\ &\delta(\vec{P}) = \delta(P_1)\delta(P_2). \end{aligned}$$

Thus, we have

$$\begin{aligned} \int_{\Gamma_0} n_1\,\mathrm{d}s = \int_{\vec{P}=0} n_1\,\mathrm{d}\sigma &= \int \left|\mathrm{D}\vec{P}\right| \delta(P_1)\delta(P_2) n_1(x,z)\,\mathrm{d}x\,\mathrm{d}z \\ &= \int \left|\mathrm{D}\vec{P}\right| \delta(\langle x,\nu\rangle - p)\delta(z - \varphi(|x|, r)) n_1(x,z)\,\mathrm{d}x\,\mathrm{d}z \qquad (3.3.5) \\ &= \int \left|\mathrm{D}\vec{P}\right| \delta(\langle x,\nu\rangle - p) n_1(x, \varphi(|x|, r))\,\mathrm{d}x. \end{aligned}$$

We can find $\left|\mathrm{D}\vec{P}\right|$ as follows:

$$\begin{aligned} &\mathrm{D}\vec{P}_1 = (\mathrm{D}_x P_1, \mathrm{D}_z P_1) = (\nu, 0), \\ &\mathrm{D}P_2 = (\mathrm{D}_x P_2, \mathrm{D}_z P_2) = \left(-\varphi'_{|x|}(|x|, r)\frac{x}{|x|}, 1\right), \\ &\left|\mathrm{D}\vec{P}\right|^2 = \begin{vmatrix} \langle \mathrm{D}P_1, \mathrm{D}P_1\rangle & \langle \mathrm{D}P_1, \mathrm{D}P_2\rangle \\ \langle \mathrm{D}P_2, \mathrm{D}P_1\rangle & \langle \mathrm{D}P_2, \mathrm{D}P_2\rangle \end{vmatrix} \\ &\quad = 1 + (\varphi'_{|x|})^2\left(1 - \frac{p^2}{|x|^2}\right) \\ &\quad = 1 + (\varphi'_{|x|})^2\frac{x^2}{|x|^2} - (\varphi'_{|x|})^2\frac{p^2}{|x|^2} = 1 + (\varphi'_{|x|})^2 - (\varphi'_{|x|})^2\frac{p^2}{|x|^2}. \end{aligned}$$

We suppose that

$$\alpha^2(|x|,p^2) = 1 + (\varphi'_{|x|})^2\Big(1 - \frac{p^2}{|x|^2}\Big) > 0, \tag{3.3.6}$$

since $p \le |x|$ on the line l.

Thus, expression (3.3.5) can be written as follows:

$$\int_{\Gamma_0} n_1 \,\mathrm{d}s = \int \alpha(|x|\,,p^2)\delta(\langle x,\nu\rangle - p)n_1(x,\varphi)\,\mathrm{d}x = f(p,\theta). \tag{3.3.7}$$

In order to obtain the inversion formula for equation (3.3.7), taking into account the inversion formula for the Radon transform, it is necessary that the weight $\alpha(|x|\,,p^2)$ (the parameter r is fixed) should admit the following factorization:

$$\alpha(|x|,p^2) = \alpha_1(|x|)\alpha_2(p^2). \tag{3.3.8}$$

Note that when we deduce formula (3.3.2), the weight

$$\alpha(|x|\,,p^2) = \sqrt{\rho^2 - p^2}\Big(z^* + \frac{A}{B}\Big)^{-1}$$

admits the above factorization with α_1 and α_2 such that

$$\alpha_1(|x|) = \Big(z^*(|x|) + \frac{A}{B}\Big)^{-1}, \qquad \alpha_2(p^2) = \sqrt{\rho^2 - p^2}.$$

Theorem. *The weight $\alpha(|x|\,,p^2)$ given by formula (3.3.6) admits factorization in the form (3.3.8) if and only if*

$$n_0(z) = (A + Bz)^{-1},$$

where A and B are determined from conditions (3.1.15).

Proof. From (3.3.7) and (3.3.8) we have

$$1 + (\varphi'_{|x|})^2\Big(1 - \frac{p^2}{|x|^2}\Big) = \alpha_1^2(|x|)\alpha_2^2(p^2),$$

or

$$|x|^2\left(1 + (\varphi'_{|x|})^2\right) = |x|^2\,\alpha_1^2(|x|)\alpha_2^2(p^2) + (\varphi'_{|x|})^2 p^2.$$

Since the left-hand side of the last equality is independent of p, it follows that $\alpha_2(\cdot)$ is a linear function:

$$|x|^2\left(1 + (\varphi'_{|x|})^2\right) = |x|^2\,\alpha_1^2(|x|)(C_1 p^2 + C_2) + (\varphi'_{|x|})^2 p^2$$

or

$$|x|^2 + (\varphi'_{|x|})^2 \, |x|^2 = |x|^2 \, \alpha_1^2(|x|) C_1 p^2 + |x|^2 \, \alpha_1^2(|x|) C_2 + (\varphi'_{|x|})^2 p^2.$$

Now we collect like terms and obtain

$$\begin{aligned} &|x|^2 + (\varphi'_{|x|})^2 \, |x|^2 = |x|^2 \, \alpha_1^2(|x|) C_2, \\ &1 + (\varphi'_{|x|})^2 = C_2 \alpha_1^2, \\ &\alpha_1^2 = (1 + (\varphi'_{|x|})^2)/C_2. \end{aligned}$$

Then we have

$$|x|^2 \alpha_1^2(|x|) C_1 + (\varphi'_{|x|})^2 = 0,$$

or

$$(\varphi'_{|x|})^2 = |x|^2 \, \alpha_1^2(|x|) C_3, \qquad C_3 = -C_1,$$

$$(\varphi'_{|x|})^2 = \frac{|x| \, (1 + (\varphi'_{|x|})^2 C_3}{C_2}.$$

Set $C_4 = C_3/C_2$. Then

$$\begin{aligned} &|x|^2 \, C_4 + (\varphi'_{|x|})^2 \, |x|^2 \, C_4 = (\varphi'_{|x|})^2, \\ &(\varphi'_{|x|})^2 = \frac{|x|^2 \, C_4}{1 - |x|^2 \, C_4} = \frac{|x|^2}{C_5 - |x|^2}, \qquad C_5 = \frac{1}{C_4}, \end{aligned}$$

or

$$\varphi' = \pm \sqrt{\frac{|x|^2}{C - |x|^2}} = \pm \frac{|x|}{\sqrt{C - |x|^2}}, \qquad C > 0.$$

Denoting $u = |x|$, we obtain

$$\varphi = \pm \int \frac{u \, \mathrm{d}u}{\sqrt{C - u^2}} = \pm \frac{1}{2} \int \frac{\mathrm{d}t}{\sqrt{1 - t}} = \pm \frac{1}{2} \int \frac{\mathrm{d}q}{\sqrt{q}} = \pm \sqrt{C - u^2} + C_0,$$

where $C_0 = \text{const}$. Because the context of the problems is such that $\varphi > 0$, we choose the plus sign before the root and, finally, we obtain

$$\varphi = \sqrt{C - |x|^2} + C_0,$$

which corresponds to (3.3.4) (see Bukhgeim, Zerkal, and Pickalov, 1983). □

We now deduce formula (3.3.1) more formally. We have

$$\begin{aligned} &P_1(x,z) = \langle x,\nu\rangle - p = 0,\\ &\mathrm{D}P_1 = (\nu, 0),\\ &\mathrm{D}P_2 = 2(x, z + A/B). \end{aligned}$$

Then we calculate $|\mathrm{D}P|^2$:

$$|\mathrm{D}P|^2 = \begin{vmatrix} |\nu|^2 & 2\langle x,\nu\rangle \\ 2\langle x,\nu\rangle & 4\,|x|^2 + 4(z + A/B)^2 \end{vmatrix} = 4(\rho^2 - p^2).$$

Since

$$z = -A/B + \sqrt{\rho^2 - |x|^2} = z^*,$$

we have

$$z^* + A/B = \sqrt{\rho^2 - |x|^2},$$

and

$$P_2(x,z) = (z + A/B)^2 - (z^* + A/B)^2 = (z + z^* + 2A/B)(z - z^*).$$

Using the property of the delta function

$$\delta(f(x)(x - x_0)) = \frac{1}{f(x_0)}\,\delta(x - x_0),$$

if $f(x_0) \neq 0$ we have

$$\delta(P_2(x,z)) = \frac{\delta(x - z^*)}{2(z^* + A/B)}.$$

Since

$$\int \delta(z - z^*) n_1(x,z)\,\mathrm{d}z = n_1(x, z^*),$$

we obtain

$$T_1 = \int |\mathrm{D}P| \delta(P_1)\delta(P_2) n_1(x,z) = \frac{2\sqrt{\rho^2 - p^2}}{2} \int \frac{n_1(x,z^*)}{z^* + A/B} \delta(P_1)\,\mathrm{d}x,$$

or

$$\frac{T_1}{\sqrt{\rho^2 - p^2}} = \int \frac{n_1(x,z^*)}{z^* + A/B} \delta(\langle x,\nu\rangle - p)\,\mathrm{d}x,$$

$$n_1(x, z^*) = \Big(z^* + \frac{A}{B}\Big) R^{-1}\Big(\frac{T_1}{\sqrt{\rho^2 - p^2}}\Big).$$

As for formula (3.3.2), we note that it can be obtained by immediate substitution of (3.3.4) into (3.3.6). Then the obtained expression is substituted into (3.3.7), and we apply R^{-1} to both sides of (3.3.7).

Deduction of formula (3.3.2) allows us to apply the technique of numerical realization of inverting the Radon transform in solving the inverse kinematic problem stated in Section 3.3 of this chapter. This technique is sufficiently well developed for solving other scientific problems, which essentially simplifies the research. The values of $T_1 = T - T_0$ were obtained in numerical experiments as a result of solution of the direct boundary-value kinematic problem following the method presented in Section 3.1 with the use of (3.1.6) for deriving T_0. Numerical realization of formula (3.3.2) requires the following procedure of gathering the projection data. Suppose we have N sources and N receivers of probe radiation. We place them as shown in Figure 3.1, where the parallel system of observation is represented. This is a system of parallel lines connecting the sources and the corresponding receivers.

The measurements of travel times T are realized for various positions of sources and receivers obtained by rotation of the system of observation about the point O (the centre of the circle and the coordinate system) counterclockwise with step

$$\Delta\theta = \pi/(M-1).$$

Here M is an integer constant that represents the number of positions of the observation system. Measurements are performed for positions characterized by the angle θ_j ($0 \le \theta_j \le \pi$) between the lines connecting the source-receiver pairs and the X_1 axis:

$$\theta_j = (j-1)\Delta\theta, \quad j = \overline{1, M}.$$

The data gathered for the angle θ_j which determines the position of observation system are said to be a projection. Note that because of the reciprocity principle the last projection ($\theta_M = \pi$) does not have to be measured. It is obtained from the first projection ($\theta_1 = 0$).

As a result of the complete data gathering, we form the projection matrix whose columns are projections. Thus, the projection matrix has dimensions $N \times M$ (N is the number of rows and M is the number of columns). The characteristic feature of the projections gathered by this method is that the first and the last elements of every projection are zeros, since the source and

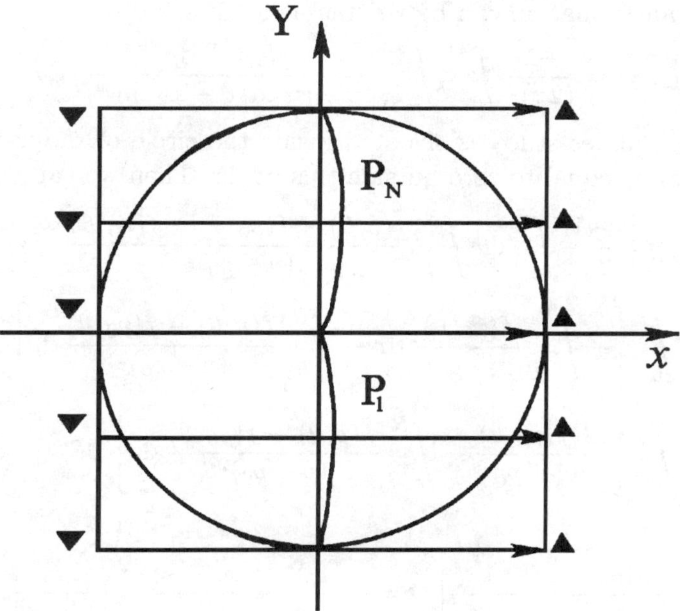

Figure 3.4: Geometrical scheme of deriving the projections for solution of the inverse kinematic problem with the use of the software package TOPAZ.

the receiver coincide in this case (see Figure 3.1). This means that actually we need two source-receiver pairs less. In the projection matrix, the first and the last rows will consist of zeros.

For numerical realization of inversion of two-dimensional Radon transform we have used the programs of the software package TOPAZ (see Bukhgeim, Zerkal, and Pickalov, 1983; Pickalov, 1985). The object under study (in this case, it is the unit disk bounded by the circle of observation system) was situated in the square $|x| \leq 1$. Outside this square the desired function (n_1) is equal to zero identically. In our case, we define the desired function to be zero in the part of the square which is situated outside the disk. In this case we can use the software package TOPAZ.

Thus, we form the projection data $F_{ij}(p_i, \theta_j)$, $i = \overline{1, N}$, $j = \overline{1, M}$, which are set in the program block of solution of the inverse problem.

The calculation of the inverse Radon transform was carried out using the algorithm of Erokhin – Shneiderov (see Erokhin and Shneiderov, 1981).

The inverse transformation can be written as follows:

$$f(x_1, x_2) = \frac{-1}{(2\pi)^2} \int_0^{2\pi} \int_{-\infty}^{\infty} \frac{\breve{f}(p,\theta)\,\mathrm{d}p}{(p - x_1\cos\theta + x_2\sin\theta)^2}\,\mathrm{d}\theta.$$

The domain under study is situated inside the circle of radius R. The desired function is equal to zero outside this circle. Then we can write

$$\begin{aligned}
f(x_1, x_2) &= \frac{-1}{2\pi^2} \int_0^{\pi} \mathrm{d}\theta \int_{-R}^{R} \frac{\breve{f}(p,\theta) - \breve{f}(p_0,\theta) + \breve{f}(p_0,\theta)}{(p-p_0)^2} \\
&= \frac{1}{2\pi^2} \int_0^{\pi} \left[-\int_{-R}^{R} \frac{\breve{f}(p_0,\theta)}{(p-p_0)^2}\,\mathrm{d}p - \int_{-R}^{R} \frac{\breve{f}(p,\theta) - \breve{f}(p_0,\theta)}{(p-p_0)^2}\,\mathrm{d}p \right] \mathrm{d}\theta \\
&= \frac{1}{2\pi^2} \int_0^{\pi} \left[\frac{2R\breve{f}(p_0,\theta)}{R^2 - p_0^2} - \int_{-R}^{R} \frac{\breve{f}(p,\theta) - \breve{f}(p_0,\theta)}{(p-p_0)^2}\,\mathrm{d}p \right] \mathrm{d}\theta, \qquad (3.3.9)
\end{aligned}$$

since

$$\int_{-R}^{R} \frac{\mathrm{d}p}{(p-p_0)^2} = \frac{1}{p-p_0}\bigg|_{-R}^{R} = \frac{1}{R-p_0} + \frac{1}{R+p_0} = \frac{2R}{R^2-p_0^2}.$$

The inner integral in formula (3.3.9) is an improper integral with singularity at the point $p_0(x)$. We take this singularity with the disk of radius $\lambda = \lambda(r)$:

$$\begin{aligned}
f(x) = \frac{1}{2\pi^2} \int_0^{\pi} \Bigg[\frac{2R\breve{f}(p_0,\theta)}{R^2 - p_0^2} - \Bigg\{ \int_{-R}^{p_0-\lambda} \frac{\breve{f}(p,\theta) - \breve{f}(p_0,\theta)}{(p-p_0)^2}\,\mathrm{d}p \\
+ \int_{p_0-\lambda}^{p_0+\lambda} \frac{\breve{f}(p,\theta) - \breve{f}(p_0,\theta)}{(p-p_0)^2}\,\mathrm{d}p + \int_{p_0+\lambda}^{R} \frac{\breve{f}(p,\theta) - \breve{f}(p_0,\theta)}{(p-p_0)^2}\,\mathrm{d}p \Bigg\} \Bigg] \mathrm{d}\theta.
\end{aligned}$$

We calculate these integrals as follows

$$\int_{-R}^{p_0-\lambda} \frac{\mathrm{d}p}{(p-p_0)^2} = \frac{1}{p-p_0}\bigg|_{-R}^{p_0-\lambda} = \frac{1}{p_0-\lambda-p_0} + \frac{1}{R+p_0} = -\frac{1}{\lambda} + \frac{1}{R+p_0},$$

$$\int_{p_0+\lambda}^{R} \frac{\mathrm{d}p}{(p-p_0)^2} = \frac{1}{p-p_0}\bigg|_{p_0+\lambda}^{R} = \frac{1}{R-p_0} - \frac{1}{p_0+\lambda-p_0} = -\frac{1}{\lambda} + \frac{1}{R-p_0}.$$

Finally, we obtain

$$\begin{aligned}
f(x) = \frac{1}{2\pi^2} \int_0^{\pi} \Bigg[\frac{2\breve{f}(p_0,\theta)}{\lambda} - \Bigg\{ \int_{-R}^{p_0-\lambda} \frac{\breve{f}(p,\theta)}{(p-p_0)^2}\,\mathrm{d}p + \int_{p_0+\lambda}^{R} \frac{\breve{f}(p,\theta)}{(p-p_0)^2}\,\mathrm{d}p \\
+ \int_{p_0-\lambda}^{p_0+\lambda} \frac{\breve{f}(p,\theta) - \breve{f}(p_0,\theta)}{(p-p_0)^2}\,\mathrm{d}p \Bigg\} \Bigg] \mathrm{d}\theta.
\end{aligned}$$

When calculating the values of $f(x)$, double numerical integration is realized using the quadratic formula of rectangles. The circuit of integration containing the singularity in the inner integral has radius $\lambda = 10^{-4}R$, which was found experimentally (see Mishen'kina, Shelud'ko, and Krylov, 1983; Mishen'kin, Mishen'kina, and Shelud'ko, 1983). The noise stability of the algorithm was studied by introducing into $f_{ij}(p_i, \theta_j)$ the errors of normally distributed pseudo-random noise proportional to the local value of $f_{ij}(p_i, \theta_j)$. Since the problem is ill-posed and the initial data contain errors, it is necessary to use the regularization process in order to solve it. Using the *a priori* information on the smoothness of the vector of the measured data and the solution of the inverse problem, we can carry out the regularization by smoothing the initial projections by cubic splines, taking into account the noise level in the initial data (see Pickalov, 1985).

Remark. It should be noted that the use of a parallel system of gathering the tomography information in model experiments is stipulated by forming the projection data in the software package TOPAZ, which was used for numerical inversion of the Radon transform. The algorithm of solving the inverse kinematic problem by the method of computerized tomography does not rule out the use of the fan-shaped scheme of projection measurement. This scheme is more simple for practical application. The similar scheme is used for forming the projection when solving the problems of industrial defectoscopy (for details see Appendix).

3.4. MODEL EXPERIMENT AND NUMERICAL INVESTIGATION OF THE ALGORITHM

Numerical experiments make it possible to evaluate the efficiency of the inversion formula (3.3.2) and demonstrate the efficiency of the tomographic approach to solution of the inverse kinematic problem. The results that are mathematically well-based are often not convenient for practical realization on computers. Thus, practical significance of the algorithm must be verified when we test it in the model experiments. Application of the algorithms which are not tested on computers often leads to doubtful results. This is an important consideration for inverse geophysical problems, since inaccurate solution of this problems can be rather costly.

Concerning the numerical investigation of the algorithm of solution of the inverse kinematic problem, it should be noted that the direct problem was solved in the complete nonlinear formulation independently of the linearized problem. This allows us to study the effect of linearization in solving the inverse problem. However, in practice, the values of T are determined from the equation

$$T = \int_{\Gamma_0} n \, \mathrm{d}s$$

resulting from the linearization of the problem. This leads to the vanishing of the error caused by linearization. So, the study of noise stability and accuracy of the algorithm is reduced to the study of the stability of the inverse operator for deriving n_1. Thus, we have pseudo-improvement of the solution stability.

In this section we give the results of numerical experiments performed on a computer. These are the following:

1) solution of the direct kinematic problem with the use of the methods stated above;
2) solution of the inverse kinematic problem with the use of the results of Sections 3.2, 3.3;
3) the study of noise stability on the basis of the obtained results and the analysis of the possibility of applying the regularization process;
4) estimates of the parameters of the observation system.

The research was done for the models of two types:

a) the quadratic term,

$$n_1(\bar{x}) = \omega_0 |\bar{x}|^2, \qquad \bar{x} \in \mathbb{R}^3; \tag{3.4.1}$$

b) the term of the Gaussian distribution,

$$n_1(\bar{x}) = \omega_0 \mathrm{e}^{-(\langle \omega, (\bar{x}_0 - \bar{x}) \rangle)^2}. \tag{3.4.2}$$

Here $\omega_0 = \text{const}$, ω is a constant three-dimensional vector; $\langle \cdot, \cdot \rangle$ is the scalar product in $\mathbb{R}^3$; $\bar{x}_0$ represents the vertex coordinates of the Gaussian distribution. In any case, unless otherwise stated, we assume

$$n_0(x_3) = (1 + 0.5\, x_3)^{-1}. \tag{3.4.3}$$

The choice of such functions n_1 is not arbitrary. This is explained by the character of their behavior inside the domain we study. We denote this domain by D. Since outside the domain $\bar{D}$, as was shown in Section 3.3, we have $n_1(\bar{x}) \equiv 0$, it follows that on the boundary of the domain the function in question has a jump (discontinuity of the first order) with the amplitude equal to

$$l = n_1(\bar{x})\big|_{x \in \partial D} .$$

In this case, it seems to be interesting to study the reconstruction of functions with various values of the jump l.

We consider the behavior of functions (3.4.1) and (3.4.2) in the domain $\bar{D}$. Both functions have extremal point in $\bar{D}$. Function (3.4.1) achieves its minimum at the point $(0,0,0)$ and function (3.4.2) has maximum at the point $\bar{x} = \bar{x}_0$. As the argument changes from the extremal point to the boundary of $\bar{D}$, function (3.4.1) increases monotonically and achieves its maximal value in $\partial\bar{D}$. In contrast, function (3.4.2) decreases monotonically and achieves its minimal value in $\partial\bar{D}$ (as $n_1(\bar{x}) \equiv 0$ for $\bar{x} \in \mathbb{R}^3 \cup \bar{D}$). By a special choice of the parameters ω_0 and $\bar{\omega}$ we can obtain the maximal values of the same order for the calculated function for the cases a) and b). In this case, the amplitudes of the jumps on the boundary of the domain $\bar{D}$ will differ by several orders.

Because we consider the linearized formulation, it is important to study the influence of the component n_1 (relative to the main term n_0) on the solution of the inverse problem. Naturally, we expect that the less is the component (which is regulated by the parameter ω_0), the better is the quality of reconstruction of the function n_1. Moreover, the jump of the function on the boundary of the domain is of a certain interest. For this reason, we consider the medium (3.4.1) under the following conditions:

$$\max|x_1| = \max|x_2| = 1,$$

$$0 < x_3 < 1, \qquad \omega_0 = \text{const}\,.$$

The desired function has axial symmetry about the axis x_3. Without loss of information, it is convenient to analyze the behavior of the component at the points of intersection of the surface formed by the rays and the plane $x_2 = 0$. In Figures 3.5 and 3.6 we see the results of reconstruction of the function n_1 for various values of the parameter ω_0. We see from the pictures that the quality of reconstruction becomes worse with the increase in the relative value of the component. Beginning with $\omega_0 = 0.15$, the graph of the reconstructed function considerably differs from the initial graph, conserving the major part of qualitative information. In Figure 3.6b we obtain inflections close to the ends of the domain where the function achieves its maximal value ($|x_1| > 0.25$). However, the general qualitative structure of the inhomogeneity in question is conserved.

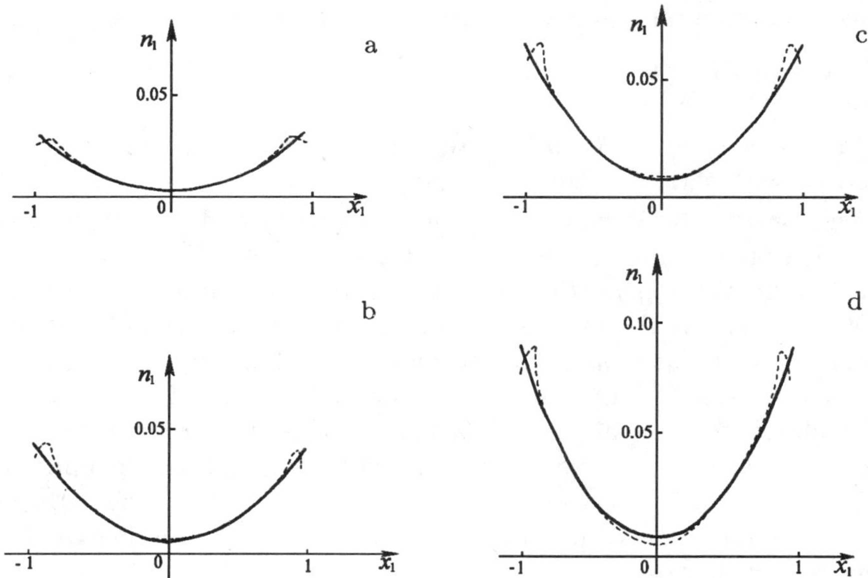

Figure 3.5: Solution of the inverse problem in the case of the quadratic component n_1: a) $\omega_0 = 0.03$, b) $\omega_0 = 0.05$, c) $\omega_0 = 0.07$, d) $\omega_0 = 0.1$. Solid line is the initial function, dashed line is the reconstructed function.

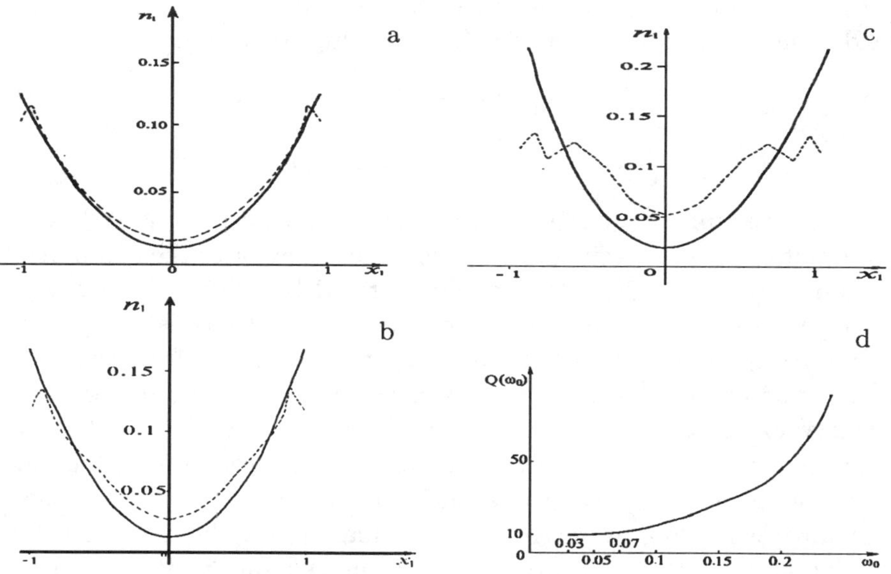

Figure 3.6: Solution of the inverse problem in the case of the quadratic component n_1: a) $\omega_0 = 0.15$, b) $\omega_0 = 0.2$, c) $\omega_0 = 0.255$, d) the graph of dependence of the error of reconstruction on the component value.

In Figure 3.6c the bends transform to oscillation. Here the distortion of the qualitative picture is seen and the information about the behavior of the function in question is lost.

The accuracy of solution of the inverse problem was defined to be the norm of deviation of the value of the function n_{1r} from the real value n_{1m}

$$Q = \sqrt{\frac{\sum_{i,j}[n_{1r}(x_{1i}, x_{2j}, x_3^*) - n_{1m}(x_{1i}, x_{2j}, x_3^*)]^2}{\sum_{i,j} n_{1m}^2(x_{1i}, x_{2j}, x_3^*)}}, \tag{3.4.4}$$

where $x_3^* = x_3^*(x_{1i}, x_{2j})$ (i and j are the current parameters of the mesh on which the values of n_{1r} are determined).

In Figure 3.6d the graph of the function $Q(\omega_0) \cdot 100\%$ is shown. We see that after $\omega_0 = 0.1$ the error function Q_0 increases rapidly. Thus, the nonlinear component can be efficiently applied for ω_0 lying within the boundaries $\omega_0 = 0.1$ and $\omega_0 = 0.15$ for the medium (3.4.1).

For the study of the noise stability of the algorithm, the noise in the initial data was produced by the generator of pseudo-random numbers. This means that the error was induced for the value of $T_1 = T - T_0$ proportionally to its local value:

$$\tilde{T}_1 = T_1(1 + z\chi/100).$$

Here z is a random value having Gaussian distribution in the interval $[-1, 1]$ and χ is the noise level in percents.

The noise level for the input data was assumed to be $\chi = 1\,\%$, $5\,\%$, $10\,\%$. In Figure 3.7a the results of reconstruction of the function (3.4.1) are represented. The analogous results for the medium (3.4.2) in the case

$$\bar{\omega} = (\sqrt{10}, \sqrt{10}, 1), \qquad \bar{x}_0 = (0, 0, -1), \qquad \omega_0 = 0.1$$

are shown in Figure 3.7b.

Figure 3.7c represents the graphs of $Q(x)$ for the medium (3.4.1) and the medium (3.4.2) in the cases a) and b), respectively. In spite of the fact that the noise level was high for the ill-posed problem, the function n_1 was reconstructed sufficiently stably. It should be noted that, in addition to the random noise, we have the following errors of systematic character in the value of T when solving the direct problem:

— the error of the Runge–Kutta method when solving the system of differential equations;

— the error of the adjustment method;

— the interpolation error.

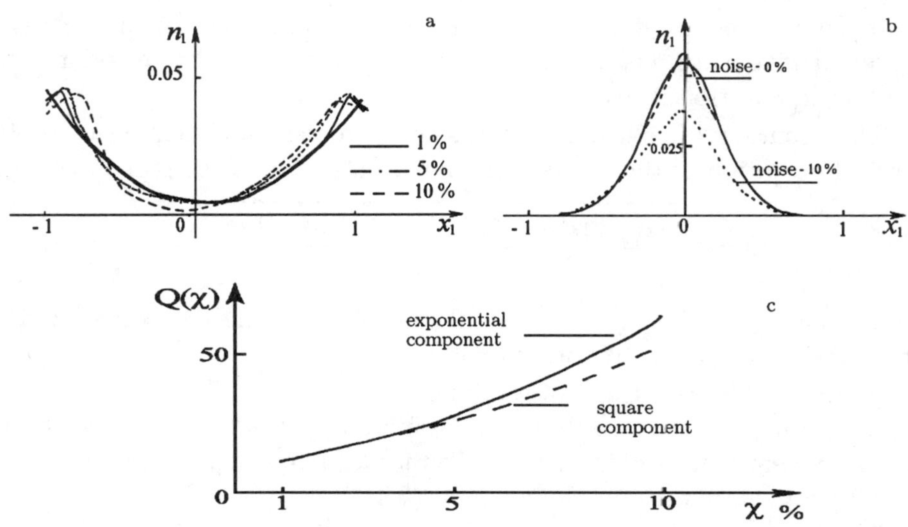

Figure 3.7: Numerical investigation of noise stability of the algorithm with respect to the random distortion of the input data: a) quadratic component, b) exponential component, c) the dependence of the error of reconstruction of n_1 on the level of noise in the initial data.

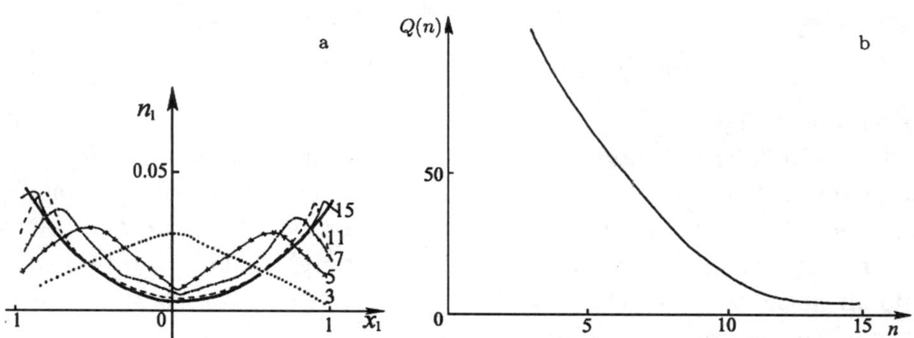

Figure 3.8: Numerical investigation of the dependence of the quality of solution of the inverse problem on the number of the source-receiver pairs used in the case of quadratic component: a) the simulation results for solution of the inverse problem; the number of the source-receiver pairs is shown in the figure; b)the dependence of the error (relative to the exact solution in percents) of reconstruction of n_1 on the number of source-receiver pairs.

Furthermore, there are the following sources of error in the algorithm of solution of the inverse problem:

— approximate character of the formula (the error of linearization);

— the inherent noise of the algorithm of inverting the Radon transform.

The systematic errors and the random noise form the total error. The difference between the results of model experiments and the exact solution is characterized by the total error. This error is, naturally, greater then the introduced noise.

The system of observation has considerable influence on the quality of reconstruction. The influence of the number of sources on reconstruction of the quadratic component is shown in Figure 3.8a. Computations were carried out for the system of observation consisting of 5, 7, 11 and 15 sources. Note that actually we have two sources less, since the first and the last one coincide with the receivers, and T at these points is equal to zero. Evidently, in the case of three sources (actually, there is one source), the reconstruction is impossible (the information is insufficient). However, for five sources, a certain qualitative results can be obtained. For seven sources, the reconstruction is better, and for 11 and 15 sources the reconstruction error is not greater than several percents in the middle of the graph. At the ends of the graph, the effect of smoothing is essential. Calculation was done for the medium with $\omega_0 = 0.05$. The domain in question was radiated from 15 directions. In Figure 3.8b we have the graph of $Q(n) \cdot 100\%$ for the quadratic component.

When we solve the tomographic problems (the inverse kinematic problem among them), it is important to reduce the number of sources radiating the probe signal and the number of directions from which the object is radiated. It is determined by both mathematical and economy reasons and by the difficulty of calculations. So, if the number of radiation sources is equal to N and the number of directions of radiation is equal to M, then the total number of measurements is $N \times M$. However, the contributions of sources and projections (directions of radiation) to the quality of the solution of the inverse problem are different. Therefore, even in the case $N = M$ the economy of one source is not equivalent to the economy of a single projection. Moreover, for specific types of media there exist upper and lower bounds N', M' and N^*, M^*, i.e., the values such that for $N < N'$ and $M < M'$ an essential part of information is lost and in the case $N > N^*$ and $M > M^*$ there is no essential improvement of solution. The results of investigation of the medium (3.4.2) illustrating the behavior of the solution of the inverse problem for small values of the numbers N and M are given below. Calculations were done for $N = 3$ and $N = 15$. The numbers of projections were $M = 3, 5, 9$. The parameters of the medium were chosen to be the following:

$$\bar{\omega} = (\sqrt{15}, \sqrt{15}, 1), \qquad \bar{x}_0 = (0, 0, -1), \qquad \omega_0 = 0.1.$$

In this case the Gaussian distribution is investigated for the structure which is thinner than in previous experiments. Figure 3.9 represents the axonometric image of the original function at the points of the surface formed by the rays stretched on the circle $x_1^2 + x_2^2 = 1$. Its projection onto the plane $x_3 = 0$ is pictured in the form of isolines. The results of solving the inverse problem are shown in the figures below.

In Figure 3.10 the case $N = 5$, $M = 3$ is represented. Figure 3.11: $N = 5$, $M = 5$; Figure 3.12: $N = 5$, $M = 9$; Figure 3.13: $N = 15$, $M = 3$; Figure 3.14: $N = 15$, $M = 5$; Figure 3.15: $N = 15$, $M = 9$. We see that, beginning with $M = 5$ and for $N = 5$ and $N = 15$, the increase in the number of projections does not lead to the improvement of the reconstruction.

The essential difference is seen only between the cases $M = 3$ and $M = 5$, which is caused by the so-called artifacts. Artifacts are represented by the wavy structures at the bottom of the Gaussian distribution. The number of artifacts is proportional to the number of projections. They are caused by the inherent noise of the algorithm of the inverse Radon transform. The number of artifacts increases as the number of directions of radiation increases. In this case, their amplitudes diminish. In Figure 3.10 ($M = 3$) the artifacts are shown by arrows. The number of artifacts is equal to 3.

If we consider the reconstruction for a fixed number of projections and different numbers of sources (for example, see Figures 3.12 and 3.15), we see that in Figure 3.12 the height of the Gaussian distribution is equal to 0.33, which is about 60 % of its real value, 0.55. Besides, the structure of the reconstructed Gaussian distribution at the bottom is considerably wider than that of the real one. The situation is different in Figure 3.15. Here the height of the Gaussian distribution is equal to 0.58, which constitutes 105 % of the real value. The form of the Gaussian distribution approximates the initial one satisfactorily. The graphs of $Q(m)$ for $N = 5$ and $N = 15$ are shown in Figure 3.16.

As was expected, a system of observation with greater number of sources and projections produces better results. However, certain qualitative characteristics useful in practice may be obtained also for the systems with smaller

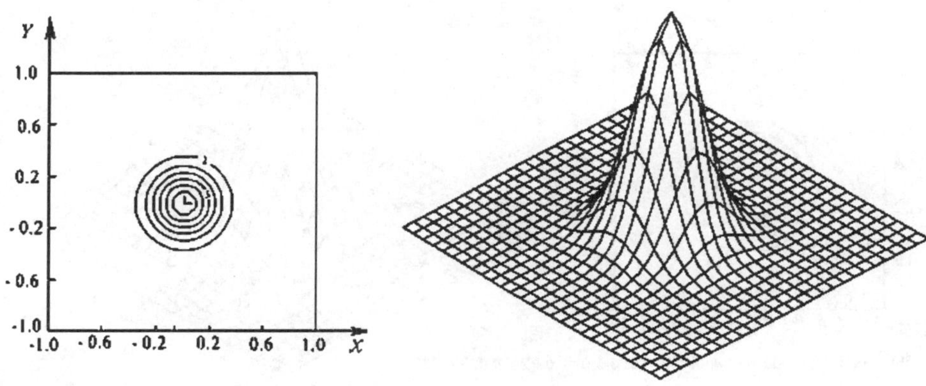

Figure 3.9

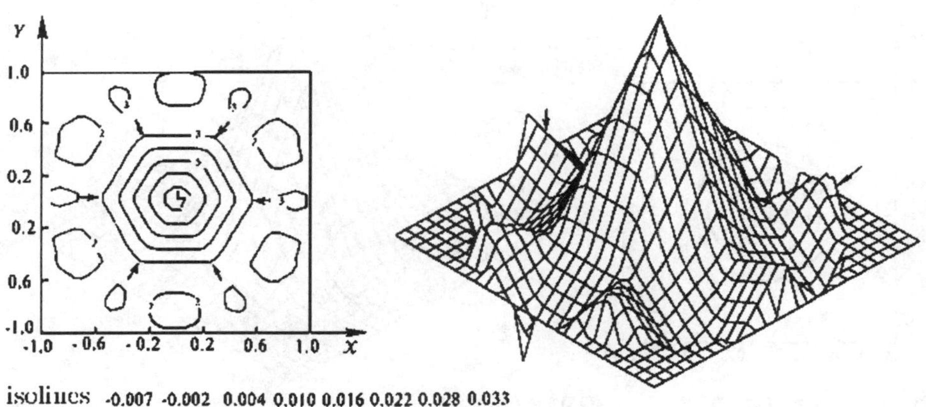

Figure 3.10

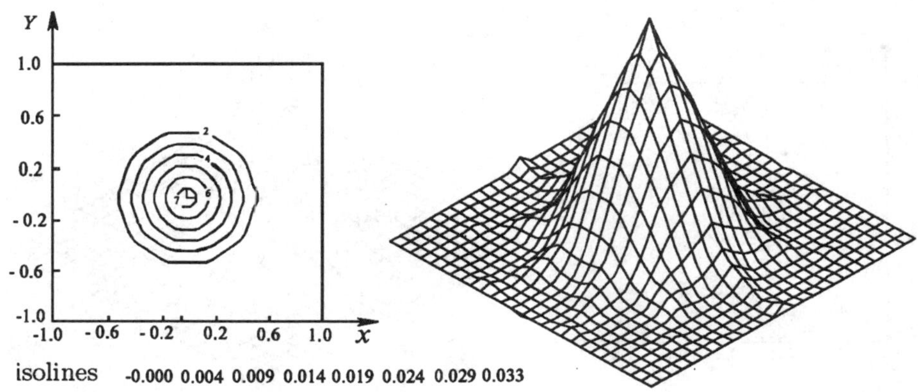

Figure 3.11

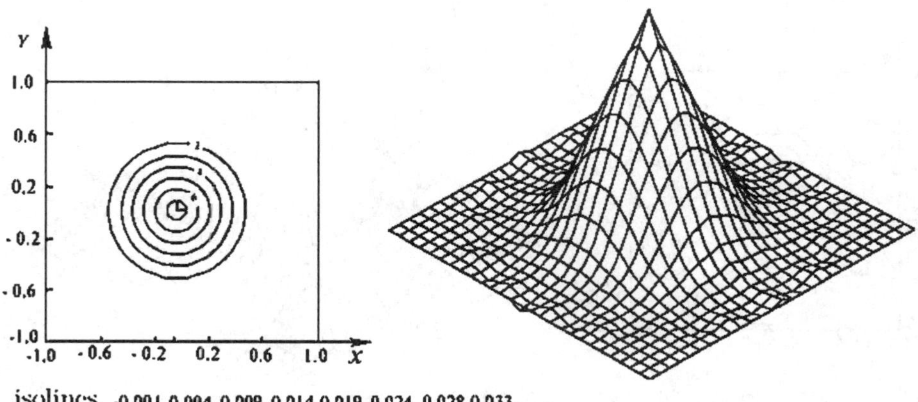

Figure 3.12

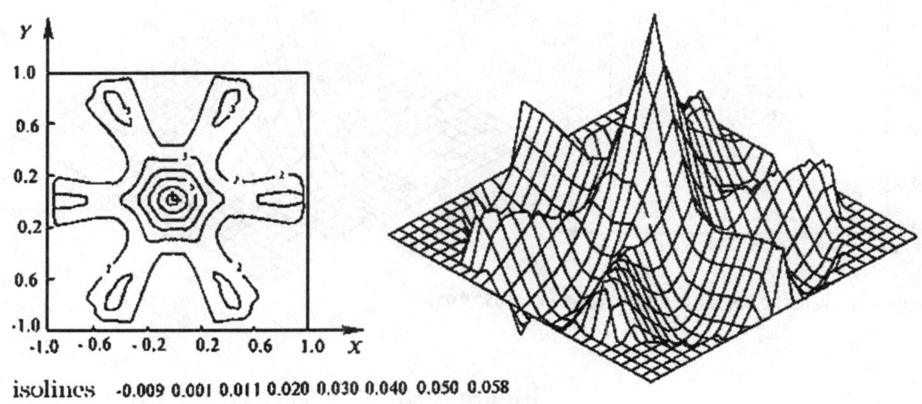

Figure 3.13

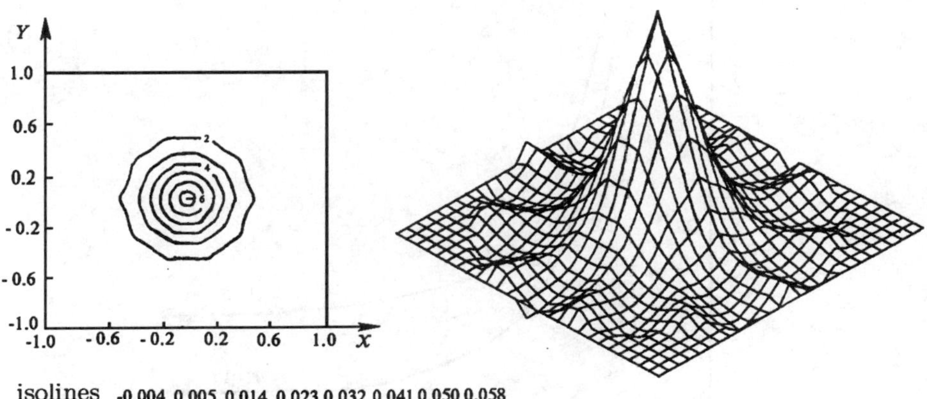

Figure 3.14

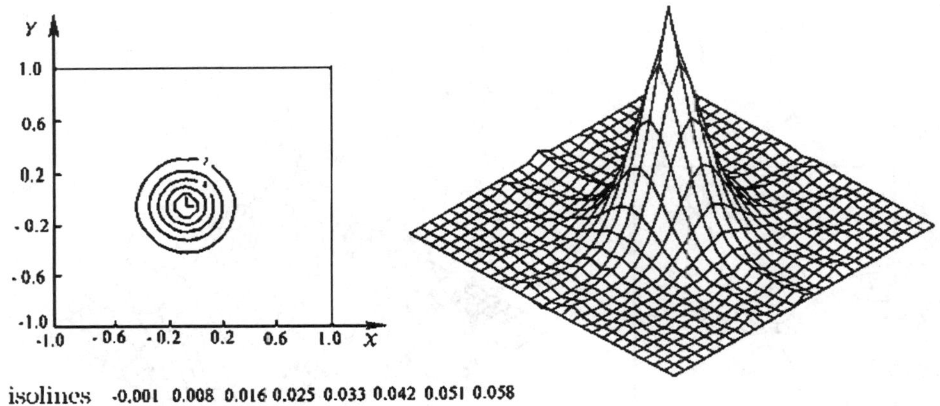

Figure 3.15

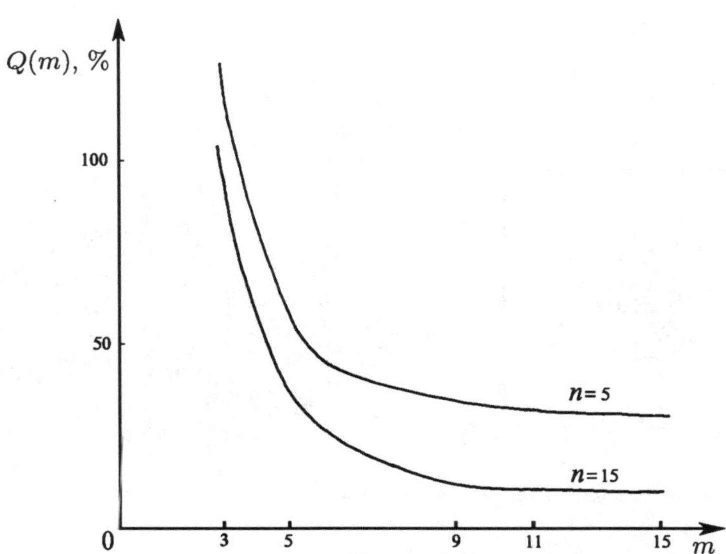

Figure 3.16: The dependence of the error of the reconstruction of the function n_1 on the number of source-receiver pairs and the number of projections.

number of sources and radiation from three directions. Now we sum up the numerical results. To find out the limits of applicability for the developed algorithm of solution of the inverse kinematic problem, we carried out the following analysis:

a) the dependence of the quality of reconstruction on the value of the component n_1 with respect to n_0 was studied;

b) using the generator of random numbers, we induced the noise to perturb the initial data, i.e., the value of $T_1 = T - T_0$, proportionally to their local values and the noise stability of the algorithm;

c) the system of observations was changed, namely, we changed the number of sources and projections.

The following conclusion can be made on the basis of a), b), and c). The upper bound of $n_1 = n - n_0$ is about 20% of n_0, when the inversion formula gives satisfactory results. The developed algorithm is stable for random noise in the input data. The parameters of the system of observation, namely, the number of sources and directions from which the studied object is radiated, affect the quality of solution of the inverse problem. We have studied the behavior of the solution of the inverse problem for small number of sources and projections. We have found out that useful information can be obtained for five sources and three directions of radiation. The quality of reconstruction of a function with a small jump on the boundary of the domain is better then for a function with large amplitude of discontinuity. The influence of sources on the quality of reconstruction is different from that of projections. The main influence of projections is the suppression of artifacts, whereas the sources are responsible for the reconstruction of inhomogeneity. This must be taken into account when performing the experiment.

In what follows we shall consider numerical experiments for determination of the index of refraction for the media where the function $n_0(x_3)$, in contrast to the cases considered above, is unknown. We have a certain initial approximation of this function determined from *a priori* information about the medium. In this case, we must solve a more general problem. Given the results of solution of the direct problem for the given medium (which are the results of physical measurements), it is required to approximate the index of refraction by the sum $n_0(x_3) + n_1(\bar{x})$. In this case, as was shown above, the ratio n_1/n_0 should be less than 0.15.

The simulation for this problem was done for the medium considered in Section 3.1, where

$$n(\bar{x}) = \sqrt{1 - x_3 - 2 \cdot 10^{-4}(x_1 + x_2) + 9 \cdot 10^{-2}\,|\bar{x}|^2}. \qquad (3.4.5)$$

The use of the explicit formula for solution of the direct kinematic problem in this case provides high-speed computation on a computer.

Formula (3.4.5) does not provide decomposition of $n(\bar{x})$ into linear component $n_0(x_3)$ and the small nonlinear component $n_1(\bar{x})$. However, because $0 \le x_3 < 1$, $|x_1| \le 1$, and $|x_2| \le 1$, formula (3.4.5) can be written as follows:

$$n(\bar{x}) = \sqrt{1 + \varphi(\bar{x})}, \qquad \varphi(\bar{x}) \ll 1, \qquad (3.4.6)$$

where

$$\varphi(\bar{x}) = -x_3 - 2 \cdot 10^{-4}(x_1 + x_2) + 9 \cdot 10^{-2}|\bar{x}|^2.$$

Furthermore, we have

$$V(\bar{x}) = \frac{1}{n(x)} = \frac{1}{\sqrt{1 + \varphi(\bar{x})}} \approx 1 - \frac{\varphi(\bar{x})}{2} = 1 + 0.5\,x_3 + V_1(\bar{x}).$$

Here

$$V_1(\bar{x}) = -(x_1 + x_2) \cdot 10^{-4} + 4.5 \cdot 10^{-2}|\bar{x}|^2.$$

Thus, the initial approximation for $n_0(x_3) = (A + Bx_3)^{-1}$ can be chosen to be $\hat{n}_0 = (1 + 0.5\,x_3)^{-1}$; condition (3.4.6) ensures that $n_1(\bar{x})$ is relatively small. Now we calculate the values T_0 for the chosen $\hat{n}_0(x_3)$. To this end, we set $T_1 = T - T_0$ and solve the inverse kinematic problem by the developed algorithm. In Figure 3.17b we have the sections of the exact function $n_1 = n - n_0$ (solid line) produced by the plane $x_2 = 0$ and the reconstructed function n_1 (dashed line). The level of the random noise having normal distribution is about 1% of the local value of T_1. Analogous noise is present in all the following experiments. The deviation Q of the reconstructed function n_{1r} from the exact one, n_{1m}, was determined by formula (3.4.4) and, in our case, is equal to 28 %. In spite of the fact that the error of reconstruction is rather high, we see from the graphs that the reconstructed function is close to the original one.

We shall now try to improve the obtained solution of the inverse problem. To this end, we fix the parameter $A = 1$ and, varying the parameter B, we study the behavior of the solution of the inverse problem in the neighborhood of the initial approximation. Figures 3.17a,b represent the results of solution of the inverse problem for the cases

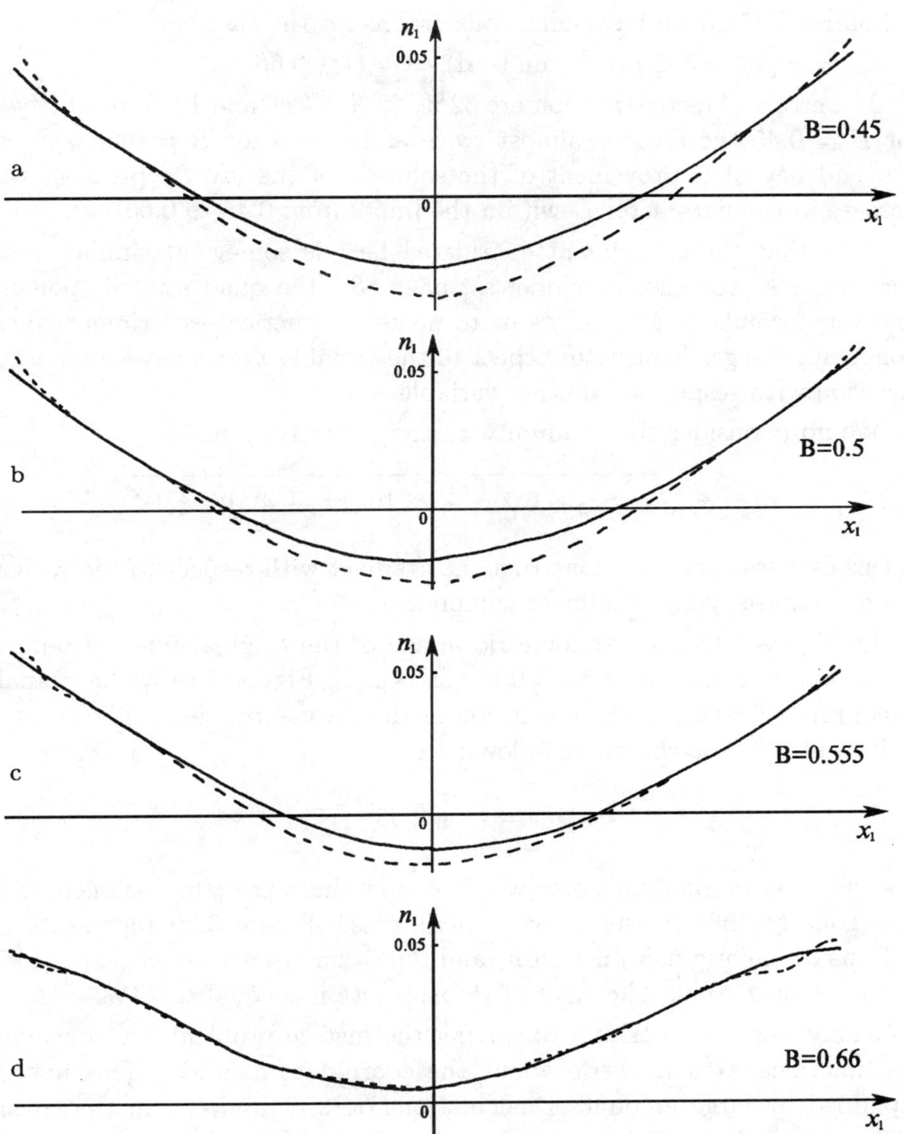

Figure 3.17: Numerical results of improving the solution of the inverse problem by exhaustive search of linear components of the speed of the signal in question.

a) $n_0 = (1 + 0.45\, x_3)^{-1}$ and b) $n_0 = (1 + 0.5\, x_3)^{-1}$.

In Figures 3.17c,d we have analogous graphs for the cases

c) $n_0 = (1 + 0.555\, x_3)^{-1}$ and d) $n_0 = (1 + 0.66\, x_3)^{-1}$.

The errors of reconstruction are 32 %, 28 %, 24 %, and 17 %, respectively. For $B = 0.45$ the error is almost twice as large as for $B = 0.66$. We see the tendency of improvement of the solution of the inverse problem with increase in the parameter B within the limits from 0.45 to 0.66.

Note that the experiment considered here is somewhat similar in the form of n_1 to the case of model a), page 80, (the quadratic component); however, formula (3.4.5) allows us to make a numerical experiment where, along with the gradient with respect to the variable x_3, we have comparable variation with respect to another variable.

We now consider the medium with the refractive index

$$n(\bar{x}) = \sqrt{1 - x_3 + 0.2\, x_1 - 2 \cdot 10^{-4} x_2 + 4 \cdot 10^{-2}\, |\bar{x}|^2}.$$

In this case we need to reconstruct the gradient with respect to the variable x_1 on the basis of the nonlinear component.

In Figure 3.18 the axonometric image of the original function and its projection onto the plane $x_3 = 0$ are shown. In Figure 3.19 we have analogous graphs for the result of solution of the inverse problem. The function n_0 in this case was chosen as follows:

$$n_0(x_3) = (1 + 0.65\, x_3)^{-1}.$$

The level of random noise was 1 % and the noise was assumed to be analogous to that in the above experiments. Figure 3.20 represents the sections of the original function n_1 and the reconstructed function produced by the plane $x_2 = 0$. The error of reconstruction is equal to 16 %.

Analyzing the algorithm of solving the inverse problem, we considered the functions n_1 symmetric about the coordinate axis x_3. This may be explained by large amount of calculations on a computer. In the considered cases it was sufficient to calculate only a half of the first projection, from which the others could be generated (see Zerkal and Khogoev, 1984; Zerkal, 1987). Below we determine the function n_1 in the case when its axis of symmetry is not passing through the centre of the coordinate system (see Zerkal and Khogoev, 1985). In this situation we need to calculate every projection separately, which increases the time requred for solving the direct problem.

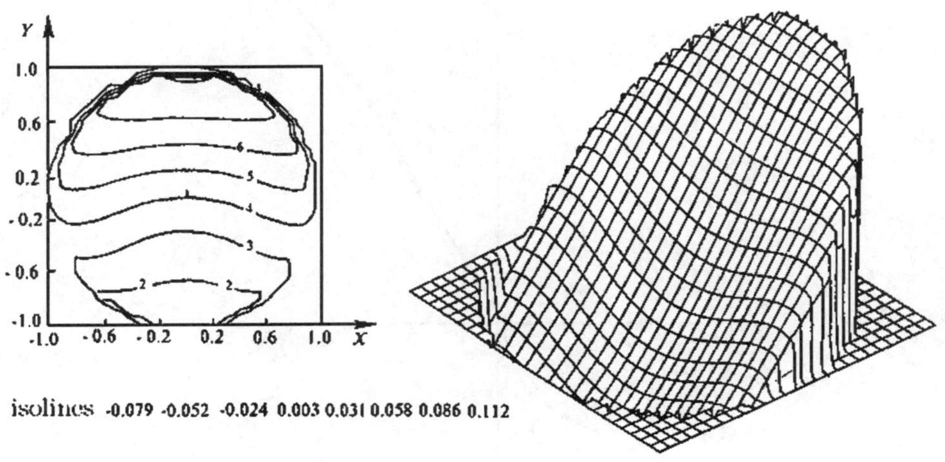

Figure 3.18

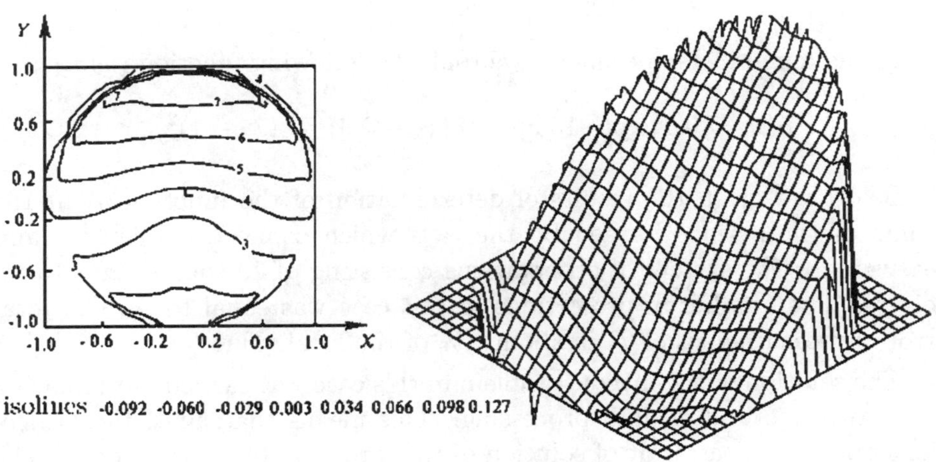

Figure 3.19

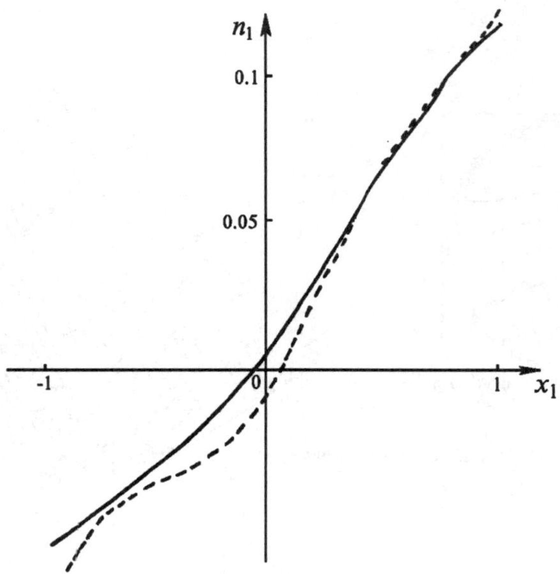

Figure 3.20: Simulation results for determination of a local perturbation of the linear index of refraction in an arbitrary part of the desired domain in three-dimensional space.

In the numerical experiment, we took the following function (as n_1):

$$n(\bar{x}) = 0.1 \exp(-15\, x_1^2 - 15(x_2 - 0.3)^2 - (x_3 - 1)^2).$$

Below are given the results of determination of the function n_1 at the points of the surface formed by the rays which connect the sources and receivers for the system of observations consisting of 15 sources and 15 receivers. The number of projections in this case was equal to 7. The total error of measurement of T_1 is about $3\,\%$ of the local value.

The solution of the inverse problem in this case was carried out following a practical method of data processing. This means that, first, the projections obtained as a result of solution of the direct problem were input. The values were approximated up to three decimal places, which led to an additional error. Then the induced projections were smoothed by cubic splines proportionally to the upper estimate of noise level.

Figure 3.21 represents the axonometric image and the projection of the original function n_1 onto the plane $x_3 = 0$. In Figure 3.22 we have similar

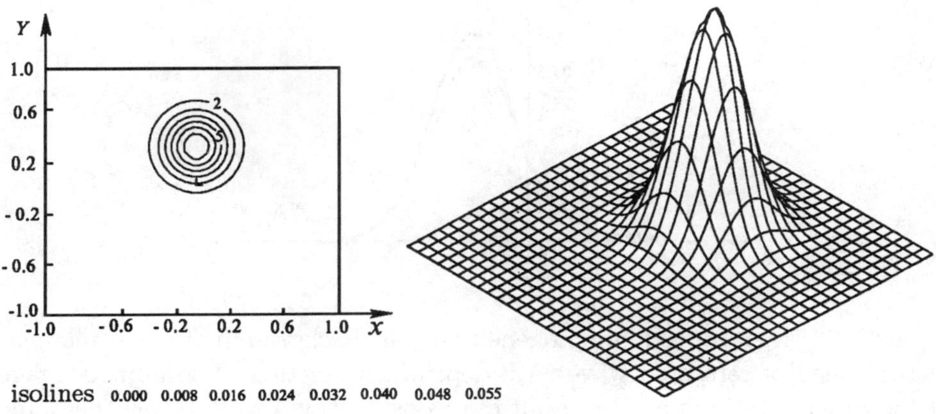

Figure 3.21

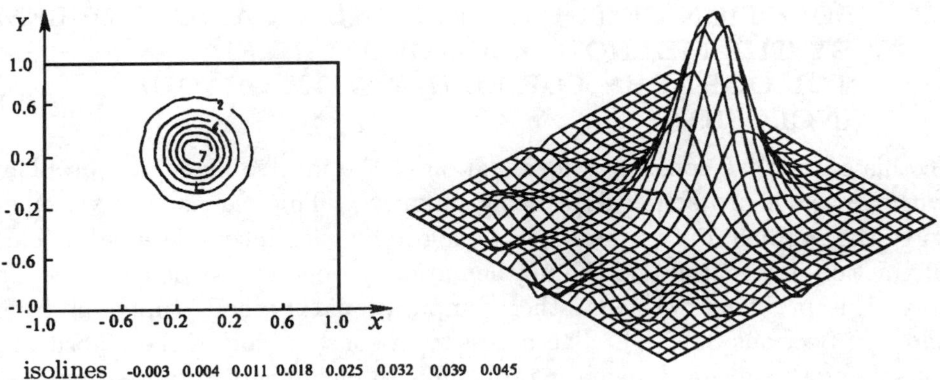

Figure 3.22

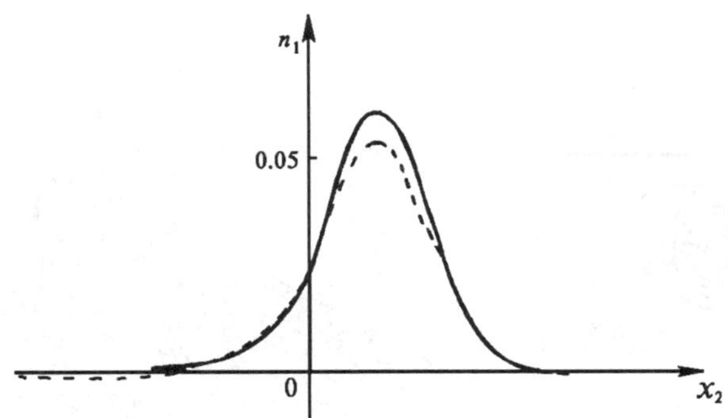

Figure 3.23: Local speed inhomogeneity on the background of linear increase in the speed of refracted wave with depth. The section of combined image of the exact solution and the result of reconstruction produced by the plane $x_1 = 0$.

graphs for the solution of the inverse problem. In Figure 3.23 the combined images of the sections of the reconstructed function and the exact function n_1 produced by the plane $x_1 = 0$ are given.

3.5. SOLUTION OF THE INVERSE KINEMATIC PROBLEM BY THE METHOD OF COMPUTERIZED TOMOGRAPHY FOR MEDIA WITH OPAQUE INCLUSIONS

In this section we consider a formulation of the inverse kinematic problem with a part of projection data being unknown. This means that we must solve the problem of computerized tomography with incomplete data. This distinction is essential and the mathematical problem must be changed. To solve this problem, we need other computational methods (different from those in Sections 3.3–3.4). The necessity of construction of the algorithms for solving the inverse kinematic problem with incomplete data is stimulated by practical needs, since some components of the vector of the measured data (projection matrix) are often unknown.

This incompleteness can be explained by two possible basic reasons:

a) in the medium under investigation there are domains opaque for probing radiation;

b) we cannot carry out the necessary measurement completely.

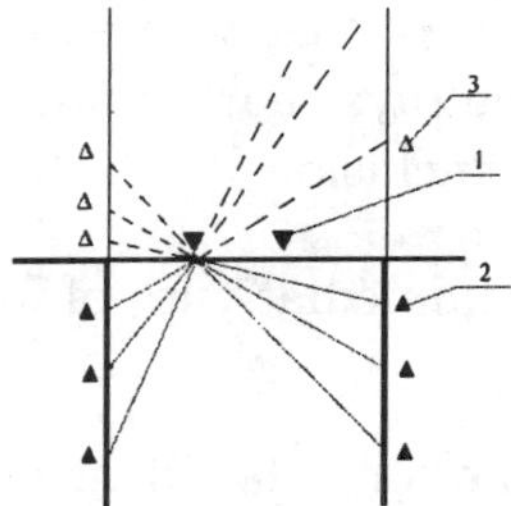

Figure 3.24: The system of observation which reduces the solution of the inverse kinematic problem to inversion of the Radon transform in a strip: 1 — sources; 2 — receivers; 3 — imaginary sources-receivers.

The situation a) arises, for example, when the medium contains absorbing objects such as wave guides, gas and oil locations, weakened zones and so on. The situation b) arises when interpreting the data of interwell radiation or processing teleseismic information.

Indeed, in the simple case of interwell research we have two parallel vertical wells where the seismoreceivers are situated. The sources are situated at the day surface ($z = 0$) in the line connecting the mouths of these wells. Thus, we obtain the system of observation in the form of the letter "П", which is not closed from below (see Figure 3.24). The reciprocity principle allows us to use the main specific feature of such a system of observation. If we remove the top of the letter П, i.e., if we model the extension of both wells upwards (dotted lines in Figure 3.24), then we obtain the system of observation consisting of two parallel lines. Thus we have obtained the problem of inverting the Radon transform in a strip. In this case, due to the shaping of the observation system to the linear form, we obtain the completeness of the projection data (Gol'din's terminology). The problem of inversion of the integral Radon transform in a strip is of interest from the viewpoint of other applied problems, in particular, for the satellite research of the Earth-orbital space. The theoretical research on this problem can be found in Zerkal and Khakhlyutin (1988), Khakhlyutin (1989). There are also inversion formulas and constructive algorithms of numerical inversion of the Radon transform in a strip for certain settings.

The problem being solved in this chapter is stated as follows. In a regular domain with the properties indicated in Section 3.3 (the formulas for the speed (3.1.14)–(3.1.15)) we take a bounded closed domain W with the property that the ray trajectories passing through it are excluded from consideration. It is required to reconstruct the index of refraction from the kinematic data T given in $p \in G$, where G is a bounded closed subset of $\mathbb{R}^1$ ($G \in \mathbb{R}^1$) containing the rays that do not pass through the domain of localization of the absorbing inclusion W.

Remark. Note that the immediate use of formula (3.2.2) is not impossible in this case.

It is impossible to use the formulas of inversion of the integral Radon transform given for $p \in \mathbb{R}^1$ (i.e., for all p such that $|p| < \infty$) in the situation when some values of $p \in \mathbb{R}^1 \setminus G$ are absent. Many of such problems have been solved in computerized tomography. Presented below is an approach to solution of the inverse kinematic problem of seismology under the conditions of local opacity of the medium which is based on reducing the integral Radon transform to the algebraic form with the subsequent application of an iterative algorithm for construction of solution using incomplete data (see Lavrent'ev *et al.*, 1995; Zerkal, 1991).

The use of iterative algorithms is explained first of all by their greater flexibility in comparison with the algorithms based on the methods of direct inversion of the integral transformation. This allows us to take into account the presence of the domains of local opacity and the *a priori* information on the object. Besides, iterative algorithms of reconstruction have great potential since they are based on fundamental domains of mathematics such as statistical analysis, linear algebra, and conditionally well-posed problems of mathematical analysis (see Bronnikov and Voskoboinikov, 1990). The inverse problem in question is related to ill-posed problems of mathematical physics, and the regularization method allows solving these problems effectively (see Bakushinskii, 1979, 1983; Bakushinskii and Goncharskii, 1989).

It is known that the iterative algebraic algorithms for reconstruction of two-dimensional distributions $\varphi(x, y)$ of physical parameters of the object in question are based on the representation of the solution in the form of the coefficients $\varphi = \{\varphi_j\}$ of the expansion

$$\varphi(x, y) = \sum_{j=1}^{M} \varphi_j \psi_j(x, y),$$

where M is a certain constant. The functions $\psi_j(x, y)$ are the basis functions for the cartesian or polar coordinate system. We denote by $f_{l,m}$ ($-L \leq l \leq L$, $1 \leq m \leq N_0$) the value of the lth ray-sum taken for the angle θ_m. Using the continuous numeration

$$i = 1 + (m-1)(2L+1) + L + 1$$

of the projection, we obtain the vector f of dimension

$$N = (2L+1)N_\theta.$$

The ith projection of this vector admits the representation of the form

$$f_i = R_i\varphi(x,y) = \sum_{j=1}^{M} \varphi_j R_i \psi_j(x,y),$$

where R_i is the operator of the integral Radon transform at the point (P_l, θ_m). Setting

$$r_{ij} = R_i \varphi_j(x,y)$$

we obtain the algebraic form of the Radon transform (see Herman, 1980)

$$R\varphi = f. \tag{3.5.1}$$

Here R is the $N \times M$ matrix with elements r_{ij}.

The choice of the basis is of great importance for reconstruction of the desired speed distribution (in our case) from (3.5.1). The basic required properties of the basis are the following:

a) linear independence;

b) approximating properties;

c) the efficiency of calculating the projections;

d) the efficiency of taking into account the *a priori* restrictions imposed on the desired distribution;

e) the quality of visual representation of the reconstructed distribution.

In Herman (1980) the bases of factorized type

$$\psi(x,y) = \tilde{\varphi}(x)\tilde{\psi}(y)$$

are proposed, in which the step functions, triangle functions, B-splines of the mth order and others can be used.

In the algorithms under consideration, most widely used is the basis of step functions on a rectangular mesh which has the properties a)–d) (see page 101) for a certain choice of the mesh spacing. The property e) is provided by the graph interpolation when constructing images.

The choice of the functions $\psi(\cdot)$ with the best approximating properties (for example, B-splines of the second and third orders) in many cases is not justified because of the complicated procedures of calculating projections and the taking into account *a priori* restrictions. Therefore, following Bronnikov and Voskoboinikov (1989), below we shall use the following procedure of building the system (3.5.1). The restriction domain (not necessarily of the rectangular form; in our case it is the disk) is divided into M pixels (a pixel is an elementary cell inside of which the properties of the medium do not change). The pixels are assumed to be squares with side Δ. We introduce the basis functions in L_2 as follows:

$$\psi_j(x,y) = \begin{cases} 1, & \text{if } (x,y) \in j\text{th pixel}, \\ 0, & \text{if } (x,y) \notin j\text{th pixel}. \end{cases}$$

Then r_{ij} is the length of the interval of the ray passing through the jth pixel and determining the ray-sum f_i, whereas φ_j is the value of the desired distribution at the jth pixel.

Remark. The matrix R corresponding to this construction procedure has two specific features:

1) even for a small number of directions, the number of its elements $N \times M \sim 10^5 \div 10^9$;

2) the number of nonzero elements is less than 10% of the total number of elements.

From this remark it follows that the procedures of direct inversion of the matrix of the system (3.5.1) (including the procedures of regularization) are not applied in practice in computerized tomography. The methods involving iterative algorithms of solving the system (3.5.1) are more effective in this case.

As was noted above, the inverse kinematic problem is related to ill-posed problems, and in order to solve it we need to construct a regularizing algorithm. Nowadays the theory and methods of construction of regularizing algorithms for linear problems are developed sufficiently well. This allows us to solve such problems effectively. However, the use of *a priori* information, for example, *a priori* restrictions imposed on the solution of the linear model (at first) leads to a nonlinear problem, which is usual for applications, particularly, in our case.

For numerical solution of nonlinear problems, the iterative approach is general. This approach for solution of ill-posed problems includes the procedure of iterative approximation of the solution of the problem. However, in contrast to the well-posed case, if the errors of measurements are present, then we cannot use the whole infinite sequence of approximations. It is necessary to stop the process. This means that we must specify a criterion of stopping the process. Thus, the iterative method of approximating the solution, the stop criterion, which corresponds to the estimate of the error level in the initial information, and the method of regularization are the elements that make up the iterative method for solving ill-posed problems (see Bakushinskii, 1979, 1983; Bakushinskii and Goncharskii, 1989). The characteristic feature of iterative algorithms is that the constructed solution belongs to a certain set Φ_A of admissible solutions. Φ_A can be defined in various ways, for example, by means of the system of inequalities

$$f_i - \delta_i \leq \langle r, \varphi \rangle \leq f_i + \delta_i, \qquad 1 \leq i \leq N.$$

Here r_i is the ith row of the matrix R, $\langle \cdot, \cdot \rangle$ is the scalar product in the M-dimensional space E^M, δ_i is the half-width of the size of the error of measurement in the projection data. Therefore, it is appropriate to use the approach where the desired solution is constructed on the basis of the variational problem

$$\inf_{\varphi \in \Phi_A} F(\varphi).$$

Here $F(\varphi)$ is a certain functional determined by the specific character of the physical problem; $\Phi_A \subset E^M$ is the set defined by the restrictions imposed on the desired vector φ.

If Φ_A consists of more than one element, then we cannot guarantee the uniqueness of the solution φ^* in the general case, since any element from Φ_A can be taken for φ^*. This fact is well known in the theory of ill-posed problems (see Tikhonov and Arsenin, 1974). In order to obtain the unique solution, it is necessary to introduce an additional rule of choosing the appropriate solution. In the optimizational approach we take φ^* as an element at which a certain functional has its minimum. As a rule, φ^* is the element with minimal square of the norm in E^M.

In general, iterative algorithms admit universal representation in the form of the one-step iterative process

$$\varphi^{k+1} = P_k T_k(\varphi^k), \tag{3.5.2}$$

where P_k are the projectors constructively determined by the set Φ_A and $T_k(\varphi^k)$ is a certain basic iterative procedure for $\Phi_A = E^M$. Evidently, the convergence of procedure (3.5.2) depends on the construction of the operators P_k, T_k and the choice of their parameters.

Note that from the point of view of representation (3.5.2) the variety of iterative algorithms is determined by the choice of the functional $F(\varphi)$, the set Φ_A and the construction of the basic procedure $T_k(\varphi^k)$.

Without dwelling on the survey and analysis of the known iterative algorithms which can be found in the special literature, concerning the problems similar to the one considered in Bronnikov and Voskoboinikov (1989), for the basic iterative procedure we take the algorithm with successive consideration of the rows of the matrix R. This algorithm allows us to find the solution of the variational problem

$$\inf \frac{1}{2}\|\varphi\|^2$$

with the restrictions

$$\underset{\sim}{f_i} \leq \langle r_i, \varphi\rangle \leq \underset{\approx}{f_i},$$

where $\underset{\sim}{f_i}$, $\underset{\approx}{f_i}$ are the upper and lower bounds for the ray-sum $\langle r_i, \varphi\rangle$.

Since the minimizing functional is positive, φ^* is unique and we use the following iterative procedure to find this element:

$$\varphi^{(k+1)} = \varphi^{(k)} + c^{(k)} r_{i(k)},$$
$$u^{(k+1)} = u^{(k)} + c^{(k)} e_{i(k)}.$$

Here $i(k)$ is the number of the row processed at the $(k+1)$th iteration; $e_{i(k)}$ is the N-component vector whose $i(k)$th element is equal to 1 and the remaining elements are equal to 0; $\varphi^{(0)} = 0$; $u^{(0)} = 0$;

$$c^{(k)} = \operatorname{mid}\left\{u_i^{(k)}, \frac{\underset{\sim}{f_i} - \langle r_i, \varphi^{(k)}\rangle}{\|r_i\|^2}, \frac{\underset{\approx}{f_i} - \langle r_i, \varphi^{(k)}\rangle}{\|r_i\|^2}\right\}.$$

Here $\operatorname{mid}\{a, b, c\}$ denotes the mean value of a, b, and c. We can show that $\varphi^{(k)}$ converges to the normal solution of system (3.5.1) if $\|\underset{\sim}{f} - f\| \to 0$ and $\|\underset{\approx}{f} - f\| \to 0$. This means that this algorithm is a regularizing one (see Bronnikov and Voskoboinikov, 1989).

The algorithm with successive consideration of the rows admits various ways of choosing the current number of the row for the kth iteration. It is usual practice to use the following simple rule, which is called the periodic setting of the index $i(k)$:

$$i(k) = k - NE(k/N).$$

This means that we successively sort the rows: for the first iteration, $k = 1$, we have

$$\varphi^{(2)} = \varphi^{(1)} + c^{(1)} r_1,$$
$$u^{(2)} = u^{(1)} + c^{(1)} e_1.$$

For the Nth iteration we have

$$\varphi^{(N+1)} = \varphi^{(N)} + c^{(N)} r_N,$$
$$u^{(N+1)} = u^{(1)} + c^{(N)} e_N,$$

and then we again take the first row, and so on.

The reconstruction algorithms involve not only the construction of the linear estimate

$$\hat{\varphi} = \varphi_\Gamma + \xi \tag{3.5.3}$$

on the basis of a certain known iterative algorithm, where φ_Γ is the exact solution and ξ is the vector of reconstruction error caused by artifacts of reconstruction and errors in the projection data. In order to obtain qualitative results, it is necessary to use certain methods of suppression of the vector ξ (depending on the complexity of the problem).

In Bronnikov and Voskoboinikov (1989) a numerical algorithm is presented which shows the efficiency of the above algorithm of examining an object with local opacity. We shall consider the method of calculation following this algorithm in application to the inverse kinematic problem with incomplete data. We shall also represent a model experiment illustrating the numerical solution of this problem.

The difficulty of solving such a problem is that shadow zones arise near the boundary of a convex opaque domain. Information may come from these zones only along certain directions. For iterative algorithms, this implies the appearance of artifacts of the step type. The reason is that the components of the vector φ corresponding to different zones are updated different numbers of times for a single iteration number. The problem of artifacts and the loss of quality of distribution in the neighborhood of an opaque region is connected also with discretization of the reconstruction domain and with the choice of parameters of the observation system. For example, the shadow zones may have considerable size and their surface can be greater than one pixel for small number of directions.

In order to guarantee the reliability of reconstruction, we propose a certain way of choosing the parameters of the observation system (see

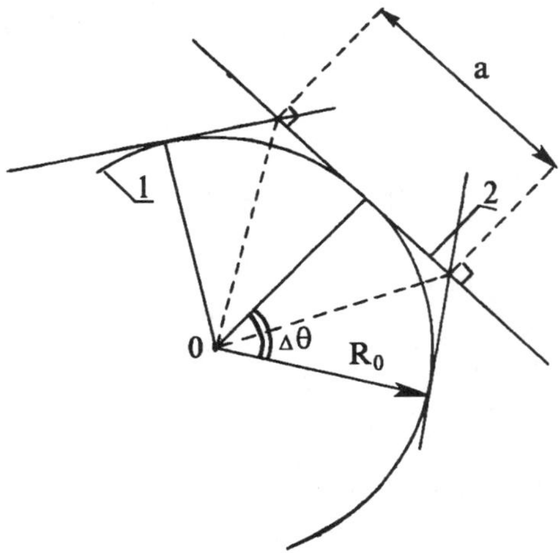

Figure 3.25: The illustration to the experiment planning: 1 — the boundary of the opaque domain; 2 — the trajectories of the rays corresponding to two nearest directions.

Lavrent'ev *et al.*, 1995). Evidently, the criteria for choosing parameters of the system of observations will depend both on the methods of registration of projections (fan-shaped or parallel system) and on the geometric shape of the opaque domains. We suppose the following:

a) the system of observations has parallel scheme of projection scanning; an image pixel is a square with side Δ equal to the step of discretization of the projection data;

b) the opaque domain is a convex one-connected domain which contains a circle with minimal radius R_0 defined by *a priori* considerations.

The projection data f_{lm} are said to have the multiplicity coefficient q if every pixel is involved in forming at least q ray-sums belonging to different directions. We now deduce the relation connecting the number of directions N_θ with the values q, R_0, and Δ. We denote by $\Delta\theta = \pi/N_\theta$ the step of discretization with respect to the angle. The value a (Figure 3.25) can be chosen as follows:

$$a = 2R_0 \tan(\Delta\theta/2).$$

We assume that the criterion of reliability is $a \leq \Delta$. In this case, we introduce the multiplicity coefficient into the last relation and take into account that the pixel with side $a = \Delta$ is intersected by at most three rays for several directions. Then we write the finite expression for the number of directions which provides the desired multiplicity of the projection data:

$$N_\theta \geq q \frac{\pi}{6 \arctan(\Delta/(2R_0))}. \tag{3.5.4}$$

Remark. If the number N_θ is fixed, for example, out of practical considerations, then the expression (3.5.4) allows estimating the pixel size corresponding to the chosen multiplicity q. This means that we can estimate the minimal surface on which the averaging of the reconstructed speed distribution will be performed.

The reconstruction algorithms, including the basic algorithm, have some essential disadvantages which hold for small numbers of directions and raysums in the projection. In the reconstructed distributions, as a rule, we have artifacts mentioned in the description of numerical experiments of Section 3.4 whose directions coincide with the angles of scanning of the projection data. Also, the distortion of the boundary takes place where distribution changes, and so on.

In order to avoid these disadvantages, we use the following nonlinear calculating procedure consisting of two stages (see Bronnikov and Voskoboinikov, 1989):

1. We construct the linear estimate (3.5.3) on the basis of the iterative algorithm of successive consideration of the rows of the matrix R.

2. In order to suppress the error vector ξ, we use the combined filter based on application of the median filtration (see Tukey, 1977) with subsequent averaging by means of the filter of running mean value.

We know that median filtration effectively suppresses impulse noises with conservation of the linear estimate. However, this filtration removes the quasi-white noise. At the same time, the filter estimate has opposite properties.

Therefore, we propose a filtration procedure consisting of two steps. At the first step we construct the estimate of the median filter

$$\hat{\varphi}_i^M = \operatorname{med}\{\hat{\varphi}_j, \quad j \in A_L(i)\}, \tag{3.5.5}$$

where A_L is a certain aperture of the filter (the aperture may have the form of a cross, square, disk, and so on); L is a parameter which defines the length $N_L = 2L + 1$ of the median filtration.

At the second step we construct the estimate of the running mean value

$$\varphi_i^{nf} = \text{aver}\,\{\hat{\varphi}_j^M : |\hat{\varphi}_j^M - \hat{\varphi}_i^M| \le h, \;\; j \in A_M(i)\}, \tag{3.5.6}$$

i.e., we find the average of only the values of $\hat{\varphi}_j^M$ that fall within the interval of length $N_M = 2M + 1$ with the aperture $A_M(i)$ and within the interval $[\hat{\varphi}_j^M - h, \hat{\varphi}_j^M + h]$.

For various values of h and L, the filter (3.5.5)–(3.5.6) can be transformed into

— the median filter, if $h = 0$;

— the filter of running mean value, if $L = 0$, $h \to \infty$.

Note that the choice of h plays the decisive role in conserving sharp changes of distributions in the linear estimate $\hat{\varphi}$. We give the following two ways of choosing the value of h:

1. If it is required to filter the quasi-white noise with dispersion σ^2 in the distribution $\hat{\varphi}$, then
$$h = c_\sigma \sigma,$$
where $c_\sigma \in [2, 3]$.

2. If it is required to conserve the change of the distribution with the amplitude H or greater in the estimate of $\hat{\varphi}$ without distortions, then
$$h = H - c_\sigma \sigma.$$

These recommendations are of qualitative character, since the result of filtration also depends on the parameters L and M. If the value of σ_i^2 at the point i is unknown, then we can use

$$\sigma_i^2 = M_L(\{|\hat{\varphi}_j - \hat{\varphi}_j^M|^2\})$$

as the dispersion estimate, and in this case h is chosen immediately.

The model experiment. The possibilities of using the inversion formulas are studied in Section 3.4; therefore, here we shall give only an example which illustrates the reconstruction of the index of refraction in the case of the opaque domain. Without loss of generality, in order to simplify the solution of the direct problem, we consider the case of one-dimensional function for the index of refraction.

We suppose that the half-space $z \ge 0$ of the three-dimensional space $\mathbb{R}^3$ is filled with a medium with index of refraction

$$n(z) = (1 + 0.5\,z)^{-1} + 0.1\,z^2.$$

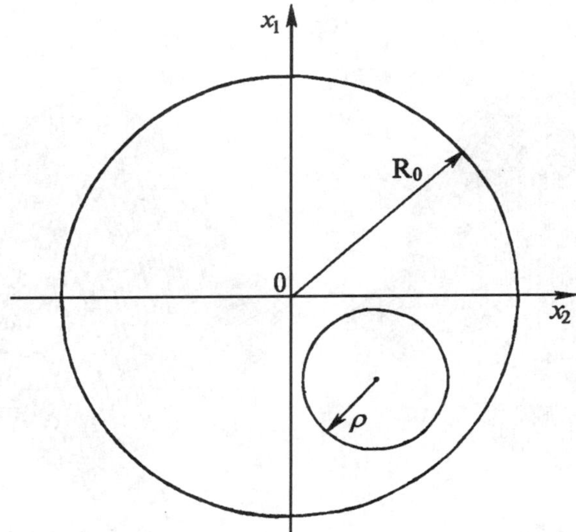

Figure 3.26: The trace of the opaque domain in the plane $z = 0$.

The opaque domain is a vertical cylinder with axis displaced from the origin (Figure 3.26). The radius of the section is 30% of the radius R_0, which coincides with the radius of the system of observation in this case.

Figure 3.27a represents the exact solution, and the result of solution of the inverse problem is shown in Figure 3.27b. The parameters taken in the experiment are as follows: $N_\theta = 16$, $R_0 = 1$, $M = 32^2$; the number of source-receiver pairs is equal to 16, $\rho = 0.3$.

In Figures 3.27c and 3.27d the real image of $n_1 = 0.1\,z^2$ and the results of solution of the inverse problem are shown. In this case, the opaque domain is a horizontal absorbing layer (the wave guide), which is shown by the cut off section of the paraboloid. $N_\theta = 16$, $M = 32^2$, the number of source-receiver pairs is equal to 16, $R_0 = 1$; the layer is situated at the depth equal to 10 % of R_0. It should be noted that though we use the approximate iteration method and approximate linearized formulation, the quality of solution of the inverse problem with incomplete data is satisfactory.

In the conclusion of the description of solving the inverse kinematic problem in the tomographic setting we note that the choice of $V_0 = A + Bz$ is of practical interest, since we know that the speed of wave propagation in the Earth grows with depth (the variable z).

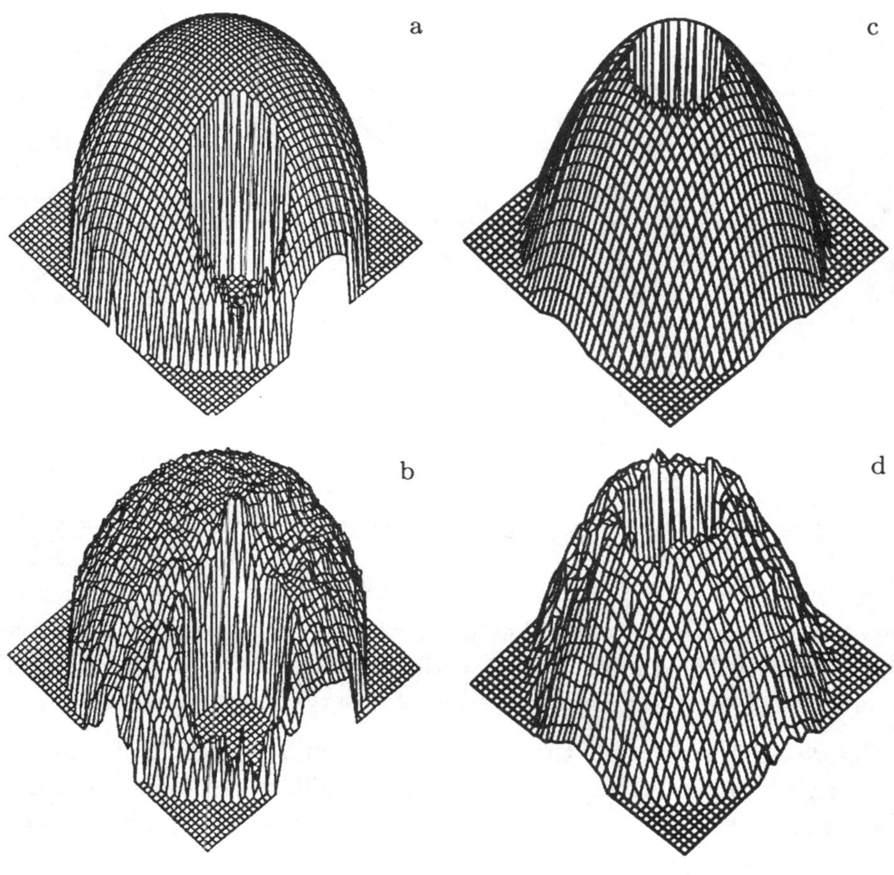

Figure 3.27: The result of numerical simulation of the solution of the inverse kinematic problem with incomplete data: a) the initial model (the exact solution); the opaque domain is a vertical cylinder, $R_0 = 1$, $\rho = 0.3$; b) the solution of the inverse problem; $N_0 = 16$, $M = 32^2$; c) the initial model (the exact solution); the opaque domain is a horizontal layer (wave guide) with underlayer depth equal to 10 % of $R_0 = 1$; d) the solution of the inverse problem; $N_0 = 16$, $M = 32^2$.

The speed of propagation of longitudinal and cross waves increases almost linearly. In most cases, in sedimentary pits which are lower than the low-speed zone, the speed grows by $0.3 \div 1$ m/s with each metre of depth (see Sharma, 1989; Sheriff and Geldart, 198)). Gas and oil accumulations can be found in deep-seated sedimentary fields with structural deformation which generates traps. When water in the porous rocks that form the sedimentary envelope is replaced by the oil or gas, the speed usually diminishes. Sometimes the speed diminishes substantially and amplitude anomalies take place because of hydrocarbon-filled regions. Oil occurs at a depth of $1.5 \div 3$ km and gas occurs at a depth of $3 \div 6$ km (see Sharma, 1989; Sheriff and Geldart, 1987). The proposed method is a version of the seismic method of refracted waves. In the case when $A = 2400$ m/s and $B = 0.5$ s^{-1} and the base is equal to 20 km (the source-receiver distance), the maximal depth for refracted waves is equal to 6.3 km (see Sharma, 1989; Sheriff and Geldart, 1987). Thus, the depth of oil and gas reservoirs is suitable for kinematic research. Hydrocarbon-filled traps may be considered as local domains of opacity. The proposed method can be successively applied for another case, namely, for the case of low-speed layers. Using the results of deep seismic probing (see Mishen'kina, Shelud'ko, and Krylov, 1983; Mishen'kin, Mishen'kina, and Shelud'ko, 1983) with the speed graphs for the earth crust in the Baikal rift zone, which has a low-speed layer, we see that the speed is easily expanded into the sum $A + Bz + v_1(x, y, z)$. Thus, the media can be studied in the considered approximation. In seismic exploration of the Earth's interior, it is especially important to explore the upper part of the the section (with depth up to 500 m) in detail because local nonhomogeneities of this part that are not taken into account can form pseudo-structures at lower structural levels. In this respect, the use of the tomographic setting considered in this chapter seems to be promising for the research in the steppe regions of the South Urals and tundra regions of the Far North.

Appendix.

Reconstruction with the use of the standard model

Intensive development of computerized tomography has led to the appearance of modern computer tomographs with unique potential that makes it possible to solve considerably more complicated problems, such as the problem of nondestructive testing of the quality of commercial products. However, it is rather difficult to provide quality processing of the testing data. It should be noted that the equipment is very complex and expensive. The volume of calculations and the amount of the measured data increase by several orders of magnitude with increase in resolution of one order of magnitude, as far as known methods are concerned.

The results obtained in industrial tomography are not so significant than those in medical tomography. Software for modern industrial tomographs in Russia is not satisfactory, whereas the specific character of the flaw detection problem requires new efficient algorithms of reconstruction.

The algorithm represented below is based on the idea of using the standard model.

1. *Problem setting and construction of the solution.* In the considered problem, X-rays are used as penetrating radiation. The source of radiation is located on the boundary of the object which is the circle of radius R. The problem is two-dimensional, the rays being the chords of the circle disposed in the form of a fan, as in Figure A.1. We denote by N the number of rays, by $\delta\varphi$ the angle distance between the neighboring rays, and by θ the fan opening, $\theta = (N-1)\delta\varphi$, $0 < \theta < \pi$. The system of observation (the source and the receivers) moves along the object boundary counterclockwise with

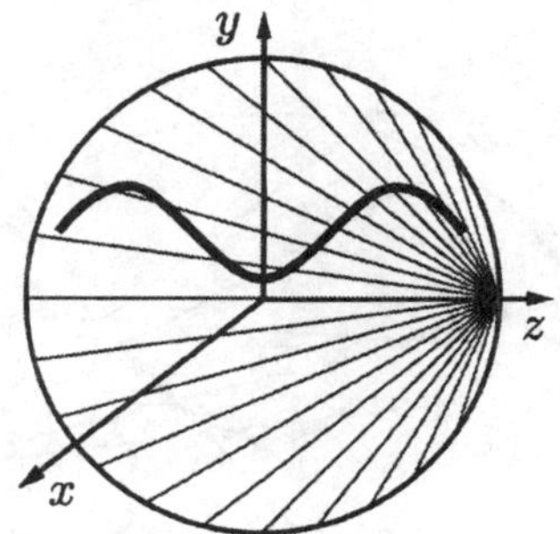

Figure A.1: The model of an item with a defect. The width of the defect is not greater than 2% of the diameter of the item. The minimum of width is attained for the ends of the defect and for the points of inflection ($R = 105$ mm). The system of observation is at the initial position, $\theta = 120°$, $N = 21$.

step $\delta\alpha = 2/M\pi$, where M is the number of positions of the source. The collection of observation data for one position of the source is said to be a projection. In order to represent the obtained projections in the form of the corresponding matrix rows (the first row is the first projection, etc.), we form the projection matrix. The number N is chosen to be even in order to avoid the loss of the ray passing through the origin.

Suppose we have the following value (ray-sum) at the receiver output:

$$P_{i,j} = \int_{l_{i,j}} \mu(x,y)\,\mathrm{d}s.$$

Here $\mathrm{d}s$ is the element of the ray $l_{i,j}$ connecting the source with the jth receiver in the ith projection; $i = 1, 2, \ldots, M$, $j = 1, 2, \ldots, N$; μ is the coefficient of absorption.

We assume that there is defect of very fine stretched structure inside the disk, and the width of the defect is substantially less than the diameter of the object. Let D denote the part of the disk which is occupied by the defect. We assume that

$$\mu(x,y) = \begin{cases} 0, & (x,y) \notin D, \\ 1, & (x,y) \in D. \end{cases} \tag{A.1}$$

This assumption allows us to simulate the "working" matrix whose elements are the differences between the corresponding elements of the projection matrix obtained for the standard model and the corresponding elements of the matrix measured for the tested object. Indeed, in the case where the

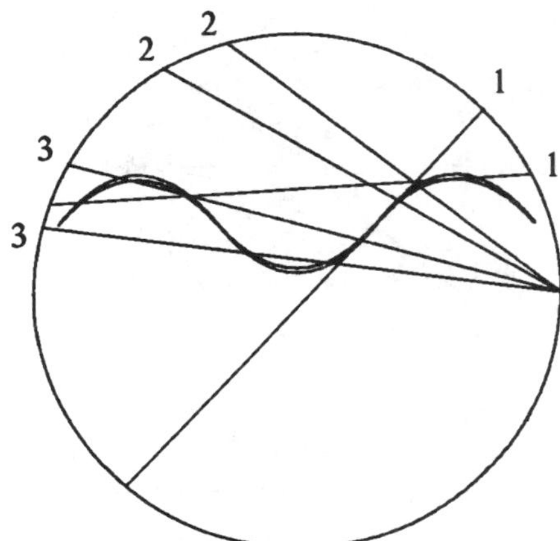

Figure A.2: The illustration to sorting a collection of rays for informativeness. The information-tangent rays are marked by 1; the ray insensitive to the defect (normal rays) are marked by 2; the noise rays are marked by 3. Let Δl_i, $i = 1,2,3$ be the lengths of corresponding ray paths within the defect, $\Delta l_1 = \max_{i,j}\{P_{i,j}\}$. Then $\Delta l_2 \ll \Delta l_1$; $\Delta l_3 < k\Delta l_1$, $0 < k < s$, where $s \in [0.1, 0.5]$.

function μ is given by formula (A.1) the nonzero elements of the working matrix are only those which correspond to the rays passing through the defect.

It is required to reconstruct the defect, i.e., to find its location and geometric form using the information about the method of obtaining the projection data and about these data in the form of the projection matrix.

From the set of rays corresponding to the nonzero values in the working matrix we select three subsets which differ by the distance of ray path within the defect. The rays that intersect the defect almost in longitudinal direction will be called "tangent" rays; the rays that ntersect the defect almost in transverse direction will be called "normal" rays. Besides, as there is no precise boundary between these subsets, there is a subset of transitional rays. It should be noted that, as the registration instruments are not sufficiently sensitive, the normal rays are absent in practice and the transitional rays are the noise.

Geometric illustration for the rays mentioned above is represented in Figure A.2. We include into consideration the defects which allow us to reduce our problem to the problem of constructing an envelope surface for the family of lines corresponding to the tangent rays. The construction of such an envelope surface in the classical sense is sufficiently difficult in practice. The method of constructing the geometric image of the defect represented below does not involve the use of the unstable operations, which require regularization.

The sorting of the collection of rays for informativeness is based on the interesting fact arising in practice from the following geometric reasoning. Let us consider two curves that are close to each other (the intersection of two close surfaces and a plane). We suppose these are concentric circles with radii R and $R+\varepsilon$, $0<\varepsilon \ll 1$. In this case, the length of the segment represented the chord Δl of the exterior circle which is tangent to the interior circle can be expressed by the following equation:

$$\Delta l = \sqrt{R^2 + 2R\varepsilon + \varepsilon^2 - R^2} = \sqrt{2R\varepsilon + \varepsilon^2}.$$

Omitting ε^2 we obtain

$$\Delta l = \sqrt{\varepsilon}\sqrt{2R}.$$

This means that among the rays intersecting the domain and disposed between the mentioned circles (curves) there are rays with length of order ε; this is essential for the sufficiently small values of $\varepsilon \sim 10^{-6} \div 10^{-8}$ m.

2. Computer calculations. We choose the maximal element in each row of the working matrix. As a result, we obtain the sequence of numbers $\{P_i\}$, where

$$P_i = \max_j \{P_{i,j}\}.$$

We transform this sequence into the vector P which will be called the working vector. Note that two numbers i and j define the corresponding ray and are related to each element of the working vector. We arrange the elements of the working vector in decreasing order and denote the obtained vector by P^1. For the elements of P^1 the following condition holds

$$P_1^1 \geq P_2^1 \geq P_3^1 \geq \ldots \geq P_{M-1}^1 \geq P_M^1.$$

For the numbers M and N sufficiently large, the jumps between the successive values P_i^1, $i = 1, 2, \ldots, M$ are not large, whereas the difference between the first and the last elements of P^1 is substantial. Note that such construction of the vector P^1, whose elements are defined by the projection number, is reduced to the one-parameter family of lines defined by the first element of the vector P^1. For the parameter we can choose, for example,

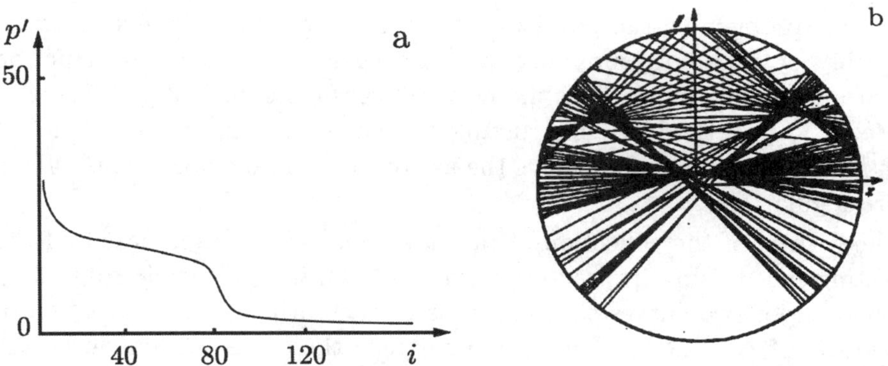

Figure A.3: The tomogram of the first degree, $N = 51$, $M = 181$, $L = 99$; a) the a-graph of the elements of the vector P^1; b) the ray image of the defect.

the polar angle of the source of radiation. The following simple method of localization of the defect can be used. Each chosen ray can be pictured with the use of numbers i and j saved in special arrays. We have to know the number of the elements of P^1 which will be used. For this purpose, we construct the graph of the elements of the vector P^1 (see Figure A.3a); then, using this graph, we visually determine the number L which is the number of the first elements of the vector P^1 that are used. Further we construct the image of the defect produced by the rays. We call this a tomogram of the first degree (see Figure A.3). Then the tomogram of the second degree is constructed, in which the image of the defect is cleared from the unnecessary parts of the rays (see Figure A.4).

Remark to the automatic definition of the number L. From the *a priori* information we know the resolution and the sensitivity threshold. This determines the quantity Δ being the minimal value of thickness (width) of a defect that can be determined by this system. In this case, the number L can be found from the estimate of the number of ray trajectories such that the values of their ray-sums satisfy the inequality

$$P_i > k\Delta, \qquad k = \text{const}, \quad i = 1, \dots, M$$

after subtraction of the corresponding value representing the model product (in our case these are the components of the vector P). The constant k is chosen from the *a priori* information including the information about geometrical conditions of measuring the values of P_{ij} and about the parameters of the defect under study.

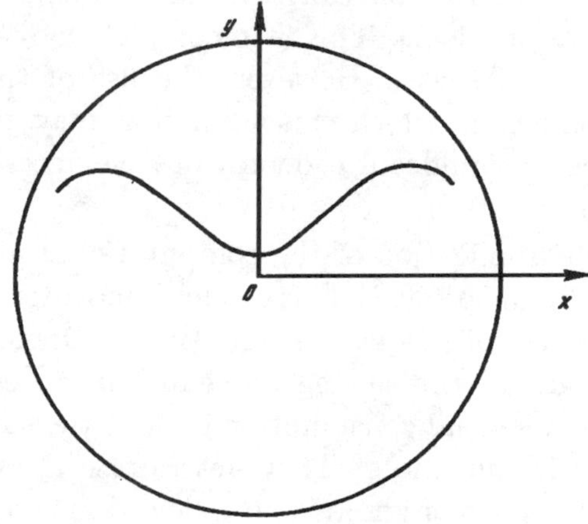

Figure A.4: The tomogram of the second degree

3. Conclusion. In conclusion we note that the proposed method makes it possible to considerably reduce the computation time required for the processing of the experimental data. This method is suitable for routine detection of long narrow defects like detachments and cracks. Unlike the known mathematical methods used in two-dimensional tomography whose potential is restricted by the necessary examination of all the sections and construction of the whole two-dimensional tomogram, the developed method allows reconstructing the most important parts or a part of the whole section as a result of their local examination.

The method of the standard model can be easily applied for the solution of the inverse kinematic problem in the tomographic setting considered in Chapter 3. In this case, a fan-shaped system of data gathering should be used for the medium model with the speed of wave propagation linearized for the speed that depends linearly on the depth. Indeed, since the projection of the rays Γ_0 onto the plane $z = 0$ (the plane of observation) in this case is represented by the chords of the corresponding circle, then it is possible to mark the contours of the inhomogeneous regions by tangent rays and obtain tomograms of the first degree. But in this case the distorted elements of the projection matrix are discarded.

Various alternative methods of solving tomographic problems are intensively developing nowadays. This also concerns the statement of the problem presented above. The problem is considered in the planar case. Naturally, three-dimensional reconstruction can be realized by successive solution of such two-dimensional problems. This is in complete agreement with tomography in the traditional sense. However, the use of the standard model method for the problems in which it is admissible may give essentially new results obtained due to significant economy of time necessary for measuring the projection data.

Another possible application of the computational method of the standard model should be mentioned. There is an important problem of detecting knots in the process of sawing tree trunks (see Danielsson *et al.*, 1997). It is required to organize the sawing procedure in an economic way from the viewpoint of the material consumption and, at the same time, to obtain the timber planks without knots. They deteriorate the quality of building materials and the presence of knots is not allowed in some joinery products and in a series of other products. For sufficiently long trunks, the time consumption of conventional algorithms of reconstruction by sections is unacceptable for the continuous production process, whereas the construction of the standard model projection matrix is not difficult because we have all the necessary *a priori* information. Thus, it becomes possible to develop more time-saving reconstruction algorithms. Furthermore, the defects to be detected are rather large in size, which allows avoiding some intrinsic difficulties of the problem of flaw detection considered above. In this case, the use of the method of the standard model is assumed to be promising.

Bibliography

Alekseev, A. S. and Tsibul'chik, G. M. (1985). Mathematical problems of seismic prospecting. In: *Modern Problems of Computational Mathematics and Mathematical Modelling.* Nauka, Novosibirsk, 91–108 (in Russian).

Alekseev, A. S. *et al.* (1969). Numerical method of solving three-dimensional inverse kinematic problem of seismology. In: *Mathematical Problems of Geophysics* (Eds. M. M. Lavrent'ev and A. S. Alekseev) Computing Centre of Siberian Branch of the USSR Acad. Sci., Novosibirsk, 179–201 (in Russian).

Anikonov, Yu. E. (1978). *Some Methods of Studying Multidimensional Inverse Problems for Differential Equations.* Nauka, Novosibirsk (in Russian).

Anikonov, Yu. E. (1973). A partial solution for one inverse kinematic problem. In: *Mathematical Problems of Geophysics* (Eds. M. M. Lavrent'ev and A. S. Alekseev). Computing Centre of Siberian Branch of the USSR Acad. Sci., Novosibirsk, 30–60 (in Russian).

Axelsson-Jacobson, C. *et al.* (1996). Comparison of three 3D reconstruction methods from cone-beam data. In: *Three-Dimensional Image Reconstruction in Radiology and Nuclear Medicine.* (Eds. P. Grangeat and J.-L. Amans). Kluwer Academic Publishers, 3–18.

Ayupova, N. B. and Golubyatnikov, V. P. (1990). Algorithms of solutions of multidimensional inverse problems and complexes of lines and planes. In: *Methods for Solution of Inverse Problems.* Institute of Mathematics, Siberian Branch of the USSR Acad. Sci., Novosibirsk, 36–44 (in Russian).

Bakushinsky, A. B. (1979). To the principle of iterative regularization. *J. Comput. Maths. Math. Phys.*, **4**, 1040–1043.

Bakushinsky, A. B. (1983). *The Principle of Iterative Regularization.* Doctorate degree thesis, Moscow State University, Moscow (in Russian).

Bakushinsky, A. B. and Goncharsky, A. V. (1989). *Ill-Posed Problems. Numerical Methods and Applications.* Moscow State University, Moscow (in Russian).

Belolipetskii, A. A. and Golov, V. I. (1989). Nonlinear algorithm of solving the inverse seismic kinematic problem in the plane. *Vestnik Mosk. Univ. Ser. 15, Vych. Mat. i Kiber.* 1, 27–29.

Belonosova, A. V. and Alekseev, A. S. (1967). On a formulation of the inverse seismic kinematic problem for two-dimensional continuously homogeneous media. In: *Several Methods and Algorithms of Interpretation of Geophysical Data*, Moscow, 137–154.

Beil'kin, G. Ya. (1978). Explicit solution of the inverse kinematic problem in the non-Herglotz case. *Vestnik nauchnogo seminara LOMI*, 78, 20–29 (in Russian).

Blagoveshchenskii, A. S. (1986). Reconstruction of a function from its integral taken along linear manifolds. *Math. Notes*, 39, 457–461.

Bronnikov, A. V. and Voskoboinikov, Yu. E. (1989). *Iterative Algorithms in the Problems of Translucent Media Tomography.* Preprint No. 18-89, Institute of Theoretical and Applied Physics, Siberian Branch of the USSR Acad. Sci., Novosibirsk (in Russian).

Bronnikov, A. V. and Voskoboinikov, Yu. E. (1990). Combine algorithms of the nonlinear filtration by the protected signals and images. *Optoelectronics, Instrumentation and Data Processing*, 1, 56–66.

Bronstein, I. N. and Semendyaev, K. A. (1964). *Handbook of Mathematics.* Van Nostrand Reinhold Company Inc., New York.

Bukhgeim, A. L., Zerkal, S. M., and Pickalov, V. V. (1983). On an algorithm of solving the inverse seismic kinematic problem. In: *Methods of Solution of Ill-Posed Problems.* Computing Centre of Siberian Branch of the USSR Acad. Sci., Novosibirsk, 38–47 (in Russian).

Censor, Y. (1983). Finite series-expansion reconstruction methods. *Proceedings of the IEEE*, 71, 409–419.

Cerveny, V., Molotkov, I. A., and Psencik, I. (1977). *Ray Method in Seismology.* Univ. Karlova, Praha.

Danielsson P. E. *et al.* (1997). Towards exact 3D-reconstruction for helical scanning of long objects. In: *Proc. Conf. Record from 1997 Int. Meeting on Fully Three Dimensional Image Reconstruction. Nemacollin, PA, June 25–28*, 1–4.

Dement'ev, V. N. *et al.* (1996). The tomographic and statistical approach for extracting anomalous structures in aerospace imagery. *Pattern Recognition and Image Analysis*, 1, 92–93.

Denisyuk, A. (1991). *The Study on Integral Geometry in Real Space.* Thesis, Moscow State University, Moscow (in Russian).

Dobretsov, N. L. (1997). The global problem of the Earth physics and geology. In: *Science on the Millennium Boundary.* Novosibirsk State University, Novosibirsk, 37–41 (in Russian).

Erokhin, V. A. and Shneiderov, V. S. (1981). *Three-Dimensional Reconstruction (Machining Tomography), Computer Simulation.* Preprint No. 23. Computing Centre of Scientific Research, Leningrad (in Russian).

Feldkamp, L. A., Davis, L. C., and Kress, J. W. (1984). Practical cone-beam algorithm. *J. Opt. Soc. Am., Ser. A.* 6, 612–619.

Finch, D. V. (1985). Cone beam reconstruction with sources on a curve. *SIAM. J. Appl. Math.*, **45**, 665–673.

Gel'fand, I. M. and Shilov, G. E. (1959). *Generalized Functions and Operations on Them.* Gos. Izd. Fiz.-Mat. Lit., Moscow (in Russian).

Gel'fand, I. M., Graev, M. I., and Vilenkin, N. Ya. (1966). *Generalized Functions, Vol. 5: Integral Geometry and Representations Theory.* Academic Press, New York – London.

Gel'fand, I. M. and Goncharov, A. B. (1987). Reconstruction of a function with compact support from its integrals over the lines intersecting a given set of points in space. *Sov. Math. Dokl.*, **34**, 373–376.

Gol'din, S. V. (1977). Certain inverse seismic kinematic problems of the reflected waves. *Dokl. Akad. Nauk SSSR*, 233, 64–67 (in Russian).

Gol'din, S. V. (1988). *Transformation and Reconstruction of the Discontinuities in the Problems of the Tomographic Type.* Institute of Geology and Geophysics, Siberian Branch of the USSR Acad. Sci., Novosibirsk (in Russian).

Gol'din, S. V. (1997). Inverse problem in radial seismic tomography. *Russian Geology and Geophysics*, 38, 1019–1037.

Grangeat, P. (1987). *Analyse d'une Systeme d'Imagerie 3D par Reconstruction a Partir de Radiographies X en Geometrie Conique.* These de doctorat, Grenoble.

Grangeat, P. (1990). Mathematical framework of cone beam 3D reconstruction in terms of the first derivative of the Radon transform. In: *Proc. Conf. Mathematical Methods in Tomography. Oberwolfach, Germany, June 5–11* (Eds. G. T. Herman, A. K. Louis, F. Natterer). Springer-Verlag, 66–97.

Helgason, S. (1980). *The Radon Transform.* Birkhäuser, Basel, Stuttgart.

Herman, G. (1980). *Image Reconstruction from Projection.* Academic Press, New York.

Herman, G. T. and Tuy, H. K. (1987). Image reconstruction from projection: an approach from mathematical analysis. In: *Basic Methods of Tomography and Inverse Problems.* (Ed. P. C. Sabatier). Adam Hilger, Bristol and Philadelphia, 3–124.

Kasyanova, S. N. and Trofimov, O. E. (1996). Computer simulation of three-dimensional tomography algorithm for a source trajectory consisting of two intersecting circles. *Optoelectronics, and Data Processing*, 6, 45–53.

Katsevich, A. I. and Ramm, A. G. (1992). Filtered back projection method for inversion of incomplete tomographic data. *Appl. Math. Lett.*, 3, 70–80.

Kazantsev, I. G. (1997a). *The Method of Weighed Projecting in Tomography.* Preprint No. 1081. Computing Center, Siberian Branch of the Russian Academy of Sciences, Novosibirsk (in Russian).

Kazantsev, I. G. (1997b). *Analytic Solution of the Radon Problem for the Finite Number of Projections.* Preprint No. 1085. Computing Center, Siberian Branch of the Russian Academy of Sciences, Novosibirsk (in Russian).

Kazantsev, I. G., Pyatkin, V. P., and Salov, G. I. (1996). The tomographic and statistical approach to detecting anomalous structures in aerospace images. *Pattern Recognition and Image Analysis*, 4, 92–93.

Khakhlyutin, V. P. (1989). *On Determination of a Function by its Integral Characteristics on Linear Manifolds and Allied Problems of Tomography.* Preprint No. 2. Institute of Mathematics, Siberian Branch of the USSR Acad. Sci., Novosibirsk (in Russian).

Kirillov, A. A. (1961). On Gel'fand's problem. *Sov. Math. Dokl.*, 2, 268–269.

Lagunova, K. F. and Omel'chenko, O. K. (1981). On a method of determination of close earthbreaks and the explicit formulas for the ray and the time. In: *Nonclassical Problems of Mathematical Physics.* Novosibirsk, 121–129.

Lavrent'ev, M. M. *et al.* (1995). Seismic tomography of the media with quasi-linear speed change containing absorbed inclusions. *Izv. RAN. Ser. Fizika Zemli*, 6, 26–31 (in Russian).

Lavrent'ev, M. M., Romanov, V. G., and Shishatskii, S. P. (1986). *Ill-Posed Problems in Mathematical Physics and Analysis.* Translations of Mathematical Monographs, 64. American Mathematical Society, Providence.

Likhachev, A. V. and Pickalov, V. V. (1998). Synthesizing algorithm of three-dimensional tomography. *Math. Modelling*, 1, 73–75.

Magnusson Seger, M. and Danielsson, P. (1999). Cone-beam tomography for logs. In: *Proc. SSAB Symposium on Image Analysis. Gothenburg, Sweden, March 9–10, 1999* (Ed. T. Gustavsson). Chalmers University of Technology, Gothenburg, 77–80.

Mishen'kina, Z. R., Shelud'ko, I. F., and Krylov, S. V. (1983). Using the linearized setting of the inverse kinematic problem for the two-dimensional fields of times $t(x, l)$ of the refracted waves. In: *Numerical Methods in Seismic Investigations.* Nauka, Novosibirsk.

Mishen'kin, B. P., Mishen'kina, Z. R., and Shelud'ko, I. F. (1983). Detailed study of the earth's crust in the Baikal rift zone by the refracted wave data. *Geology and Geophysics*, 12, 82–91.

Mukhometov, R. G. (1975). The inverse seismic kinematic problem in the plane. In: *Mathematical Problems of Geophysics* (Eds. M. M. Lavrent'ev and A. S. Alekseev). Computing Centre of Siberian Branch of the USSR Acad. Sci., Novosibirsk, 243–254 (in Russian).

Natterer, F. (1986). *The Mathematics of Computerized Tomography.* Teubner and John Wiley, Stuttgart.

Orlov, S. (1975). The theory of three-dimensional reconstruction. Reconstruction operator. *Kristallografiya*, 20 (4) (in Russian).

Palamodov, V. and Denisyuk, A. (1988). Inversion de la transformation de Radon d'apr'es des donn'ees incomplites. *CRAS*, 307, 181–183.

Pickalov, V. V. (1985) Applied software package for solving the problems of computerized tomography. *Problems of Reconstructive Tomography.* Novosibirsk, 132–135 (in Russian).

Pickalov, V. V. and Preobrazhensky, N. G. (1987). *Reconstructive Tomography in Gas Dynamics and Plasma Physics.* Nauka, Novosibirsk (in Russian).

Pickalov, V. V. and Melnikova, T. S. (1995). *Plasma Tomography.* Nauka, Novosibirsk (in Russian).

Problems of Geotomography. (1997). (Eds. A. V. Nikolaev, I. N. Galkin, and I. A. Sanina). Nauka, Moscow (in Russian).

Psencik, I. N. (1983). Solving the direct seismic problem in horizontal inhomogeneous media by the ray method. In: *Numerical Methods in Seismic Investigations.* Nauka, Novosibirsk (in Russian).

Puzyrev, N. N. (1992). *Methods of Seismic Investigations.* Nauka, Novosibirsk (in Russian).

Ramm, A. G. and Zaslavsky, A. I. (1992). Inversion of incomplete cone-beam data. *Appl. Math. Lett.*, 5, 91–94.

Rizo, P. *et al.* (1991). Comparison of two three-dimensional x-ray cone-beam-reconstruction algorithms with circular source trajectories. *J. Opt. Soc. Am. Ser. A*, 10, 1639–1648.

Romanov, V. G. (1987). *Inverse Problem of Mathematical Physics.* VNU Science Press, Utrecht.

Schaller, S., Flohr, T., and Steffen, P. (1996). A new approximate algorithm for image 15 reconstruction in cone-beam spiral CT at small cone angles. In: *Conference Record, IEEE Medical Imaging Conference, November, 1996, Anaheim, CA.* 1703–1709.

Seismic Tomography with Applications to Global Seismology and Exploration Geophysics. (1987). (Ed. T. Nolet). D. Reidel Publishing Company.

Semyanistyi, V.I. (1961). Homogeneous functions and some problems of integral geometry in a space of constant curvature. *Sov. Math. Dokl.*, 2, 59–62.

Sharma, P. (1986). *Geophysical Methods in Geology.* Elsevier Science Publishing, New York.

Sheriff, R. and Geldart, L. (1985). *Exploration Seismology.* Cambridge University Press.

Smirnov, K.K. and Trofimov, O.E. (1979). Reconstruction of a function by its integrals along the lines. In: *Computer-Based Automatization of Scientific Research.* Novosibirsk.

Smith, B.D. (1985). Image reconstruction from cone-beam projections: necessary and sufficient conditions and reconstruction methods. *IEEE Trans. Med. Image.* MI-4, 14–28.

Smith, B.D. (1990). Cone-beam tomography: recent advances and tutorial review. *Opt. Eng.*, 5, 524–534.

Three Dimensional Image Reconstruction in Radiology and Nuclear Medicine. (1995). (Eds. P. Grangeat and J.-P. Amans). Kluwer Academic Publishers.

Tikhonov, A.N. and Arsenin, V.Ya. (1977). *Solutions of Ill-posed Problems. Scripta Series in Mathematics* (Ed. F. John). Wiley & Sons, New York.

Tikhonov, A.N., Arsenin, V.Ya., and Timonov, A.A. (1987). *Mathematical Problems of Computer Tomography.* Moscow, Nauka (in Russian).

Townsend, D.W. *et al.* (1993). Performance study of 3D reconstruction algorithms for positron emission tomography. In: *International Meeting on*

Fully Three-Dimensional Image Reconstruction in Radiology and Nuclear Medicine. Abstracts. University of Utah, Salt Lake City, Utah, USA, 3–5.

Trofimov, O. E. (1991a). On the problem of reconstruction of a function of three variables from its integrals along straight lines crossing a specified curve. *Optoelectronics, Instrumentation Data Processing*, 5, 56–62.

Trofimov, O. E. (1991b). To the problem of reconstructing a function from its integrals along the straight lines crossing the specified curve. In: *Methods for Solution of Conditionally Correct Problems.* Institute of Mathematics, Siberian Branch of the Russian Academy of Sciences, Novosibirsk, 174–188 (in Russian).

Trofimov, O. E. (1992). On the Fourier transform in the sense of generalized functions of cone-beam data. In: *Ill-Posed Problems of Math. Physics and Analysis.* Novosibirsk, 119 (in Russian).

Trofimov, O. E. (1992). Numerical algorithms for three-dimensional tomography reconstruction. *Optoelectronics, Instrumentation Data Processing*, 3, 85–91.

Trofimov, O. E. (1993a). The use of inversion formulas involving distributions in constriction of numerical algorithms for three-dimensional reconstruction. *Optoelectronics, Instrumentation and Data Processing*, 2, 55–61.

Trofimov, O. E. (1993b). Fourier transform of beam data. *Optoelectronics, Instrumentation and Data Processing*, 4, 110–112.

Trofimov, O. E. (1993c). Cone beam reconstruction and Fourier transform of distributions. In: *Lecture Notes in Computer Science. Vol. 719.* Springer-Verlag, 564-571.

Trofimov, O. E. (1995). Correlation of two methods of cone-beam reconstruction. *J. System Analysis-Modeling Simulation*, 18, 169–172.

Trofimov, O. E. (1996). Determination of point objects coordinates in computer vision as a problem of nonlinear tomography. *Optoelectronics, Instrumentation and Data Processing*, 2, 65–67.

Trofimov, O. E. (1997a). Computer modeling of cone beam reconstruction. In: *Proc. Intern. Conf. Frontiers in Industrial Process Tomography II.*

Tech. Univ., Delft, The Netherlands, April 1997. Engineering Foundation, New York, 257–260.

Trofimov, O. E. (1997b). Numerical algorithms of tomography on base distributions. In: *Proc. 15th Institute Mathematics of the ACS World Congress on Scientific Computation, Modelling and Applied Mathematics. Berlin, 1997.* Wissenschaft & Technik Verlag, Berlin, 707–711.

Trofimov, O. E. and Kasyanova, S. N. (1998). The use of 3D-tomography algorithms by determination of point objects coordinates in computer vision development of measuring techniques for multiphase flows. In: *Proc. Second International Symposium on Measuring Techniques for Multiphase Flows. Beijing, China, August, 1998.* Standards Press of China, 205–209.

Trofimov, O., Kasyanova, S., and Badazhkov, D. (1999). Algorithms of 3D cone beam tomography for incomplete data. In: *Proc. First World Congress on Industrial Process Tomography. Buxton, UK, April, 1999.* 181–183.

Tukey, D. W. (1977). *Exploratory Data Analysis.* Addison-Wesley Publishing Company.

Tuy, H. K. (1983). An inverse formula for cone-beam reconstruction. *SIAM J. Appl. Math.*, 43, 546–552.

Vainikko, T. M. and Veretennikov, A. Yu. (1986). *Iteration Procedures in Ill-Posed Problems.* Nauka, Moscow (in Russian).

Voskoboinikov, Yu. E. *et al.* (1997). Nonlinear filtering algorithms used in improvement of reconstructed tomographic image quality. *Optoelectronics, Instrumentation and Data Processing*, 3, 12–15.

Wang, H. *et al.* (1993). A general cone-beam reconstruction algorithm. *IEEE Trans. Med. Image*, 3, 486–496.

Wang, H., Jaszczak, R. J., and Coleman, R. E. (1993). A mutual information evaluation of converging collimation for three-dimensional myocardial SPECT imaging. In: *Abst. Intern. Meeting on Fully Three-Dimensional Image Reconstruction in Radiology and Nuclear Medicine.* Univ. of Utah, Salt Lake City, Utah, USA, 116–117.

Zeng, G. L., Clack, R., and Gullberg, G. T. (1993). Implementation of Tuy's cone beam inversion formula. In: *Abst. Intern. Meeting on Fully Three-*